Mike Farquharson-Roberts holds a PhD in Maritime History from the University of Exeter. He previously had a long and distinguished career in the Royal Navy.

──────────

'This book is essential and enjoyable reading for anyone who wants to understand the nature of the truly global war fought beyond the trenches of the Western Front. Insightful, authoritative and digestible, it opens a window onto the Royal Navy's vital but largely unseen work: from executing the strategic economic blockade which tied a noose around the neck of the Kaiser's Germany, to the little known but significant contribution of the Royal Naval Division in the land war. It also shines a penetrating spotlight on the Royal Navy's innovation, people and organisation, as well as the unfolding revolution in its battle space, as submarines and aircraft emerged to transform warfare at sea, and from the sea.'

Admiral Sir George Zambellas KCB DSC ADC DL
First Sea Lord and Chief of Naval Staff

A History of the Royal Navy: The Age of Sail
Andrew Baines (ISBN: 978 1 78076 992 9)

A History of the Royal Navy: Air Power and British Naval Aviation
Ben Jones (ISBN: 978 1 78076 993 6)

A History of the Royal Navy: The American Revolutionary War
Martin Robson (ISBN: 978 1 78076 994 3)

A History of the Royal Navy: Empire and Imperialism
Daniel Owen Spence (ISBN: 978 1 78076 543 3)

A History of the Royal Navy: The Napoleonic Wars
Martin Robson (ISBN 978 1 78076 544 0)

A History of the Royal Navy: The Nuclear Age
Philip D.Grove (ISBN: 978 1 78076 995 0)

A History of the Royal Navy: The Royal Marines
Britt Zerbe (ISBN: 978 1 78076 765 9)

A History of the Royal Navy: The Seven Years' War
Martin Robson (ISBN: 978 1 78076 545 7)

A History of the Royal Navy: The Submarine
Duncan Redford (ISBN: 978 1 78076 546 4)

A History of the Royal Navy: The Victorian Age
Andrew Baines (ISBN: 978 1 78076 749 9)

A History of the Royal Navy: Women and the Royal Navy
Jo Stanley (ISBN: 978 1 78076 756 7)

A History of the Royal Navy: World War I
Mike Farquharson-Roberts (ISBN: 978 1 78076 838 0)

A History of the Royal Navy: World War II
Duncan Redford (ISBN: 978 1 78076 546 4)

The Royal Navy: A History Since 1900
Duncan Redford and Philip D. Grove (ISBN: 978 1 78076 782 6)

A HISTORY OF THE
ROYAL NAVY
World War I

Mike Farquharson-Roberts

in association with

Published in 2014 by I.B. Tauris & Co. Ltd
6 Salem Road, London W2 4BU
175 Fifth Avenue, New York NY 10010
www.ibtauris.com

Distributed in the United States and Canada Exclusively by Palgrave Macmillan,
175 Fifth Avenue, New York NY 10010

ISBN: 978 1 78076 838 0
eISBN: 978 0 85773 542 3

A full CIP record for this book is available from the British Library
A full CIP record is available from the Library of Congress

Library of Congress Catalog Card Number: available

Printed and bound in Great Britain by T.J. International, Padstow, Cornwall

Contents

List of Illustrations

Tables

Figures

Maps

Colour Plates

Series Foreword

The Royal Navy has for centuries played a vital if sometimes misunderstood or even at times unsung part in Britain's history. Often it has been the principal – sometimes the only – means of defending British interests around the world. In peacetime the Royal Navy carries out a multitude of tasks as part of government policy – showing the flag, or naval diplomacy as it is now often called. In wartime, as the senior service of Britain's armed forces, the navy has taken the war to the enemy, by battle, by economic blockade or by attacking hostile territory from the sea. Adversaries have changed over the centuries. Old rivals have become today's alliance partners; the types of ship, the weapons within them and the technology – the 'how' of naval combat – have also changed. But fundamentally what the navy does has not changed. It exists to serve Britain's government and its people, to protect them and their interests wherever they might be threatened in the world.

This series, through the numerous individual books within it, throws new light on almost every aspect of Britain's Royal Navy: its ships, its people, the technology, the wars and peacetime operations too, from the birth of the modern navy following the restoration of Charles II to the throne in the late seventeenth century to the war on terror in the early twenty-first century.

The series consists of three chronologically themed books covering the sailing navy from the 1660s until 1815, the navy in the nineteenth century from the end of the Napoleonic Wars, and the navy since 1900. These are complemented by a number of slightly shorter books which examine the navy's part in particular wars, such as the Seven Years' War, the American Revolution, the Napoleonic Wars, World War I, World War II and the Cold War, or particular aspects of the service: the navy and empire, the Women's Royal Naval Service, the Royal Marines, naval aviation and the submarine service. The books are standalone works in their own right, but when taken as a series present the most comprehensive and readable history of the Royal Navy.

Duncan Redford
National Museum of the Royal Navy

The role in Britain's history of the Royal Navy is all too easily and too often overlooked; this series will go a long way to redressing the balance. Anyone with an interest in British history in general or the Royal Navy in particular will find this series an invaluable and enjoyable resource.

Tim Benbow
Defence Studies Department,
King's College London at the
Defence Academy of the UK

Acknowledgements

I am very grateful to the National Museum of the Royal Navy, the Fleet Air Arm Museum and the Royal Navy Submarine Museum for making so many pictures from their archives available to me. In particular I cannot thank the various curators, archivists and keepers of pictures who have assisted me in preparing this book enough. They have invariably been extremely helpful; rather than simply deal with a request, they have gone further, providing me with a source or picture that I should have asked for as well as the one I did request. Most of the illustrations used in this book derive from the former! In purely alphabetical order, my thanks are due to Captain Richard Cosby of 'Maritime Originals', Stephen Courtney, Curator of Photographs, National Museum of the Royal Navy, Barbara Gilbert, Archivist, Fleet Air Arm Museum, Debbie Turner-West, Keeper of Photographs, Royal Navy Submarine Museum, and Stuart Wheeler, Assistant Archive and Library Manager, The Tank Museum. My special thanks to Jenny Wraight, Librarian, Royal Naval Historical Branch, for making available the Grand Fleet Battle Orders and the contemporary Signal Book, and to General von Wilcken for his assistance in German linguistic usage; for example, without his efforts I would not have known that the German battlecruiser was known as *Von Der Tann*, and not, as would be expected, *von der Tann*. I am also indebted to Admirals James Baldrick and David Cooke for advice and screening out major technical howlers.

While all of those who have provided paintings and photographs have made every effort where appropriate to contact copyright holders, they take no responsibility for any copyright implications that may arise as a result of the publication of the images they have supplied.

I have received much helpful advice, loans of out-of-print books and technical corrections from many friends and former colleagues. However, any remaining errors are mine, for not listening enough!

Mike Farquharson-Roberts

Map 1. The North Sea, including the Western Approaches

Map 2. The Mediterranean Sea

Map 3. South America

Introduction

During the century since 1914, public perceptions of World War I have gone through many changes. Immediately after the war there was a feeling of relief and thanks for a job well done. After about a decade, with the generals safely dead and unable to sue for libel, the politicians' memoirs started to appear and the war poets such as Siegfried Sassoon came to wider prominence. What followed was a 'never again' feeling of revulsion about the carnage on the Western Front. After World War II the publication of such books as *The Swordbearers* by Correlli Barnett, which was very critical of the generalship of Field Marshal Haig, continued the process, as did the 1960s satirical musical *Oh What a Lovely War*, which became a star-studded and very successful film. The culmination was the final series of *Blackadder*, a television comedy, now even used as teaching material about the war in schools, which portrayed army commanders as unfeeling incompetents. The Royal Navy's participation in World War I was spared this process; for it the indignity was worse: it was largely ignored.

As World War I approached, the armies of all sides seemed to be preparing for a somewhat more technologically advanced Napoleonic War. Wellington would have recognized the army that Britain sent to France. He approved of troops taking cover to avoid being killed. In his day they had to stand in order to be able to fire and reload their muskets, but at Waterloo he had ordered them to lie down to avoid enemy fire. He would only have seen as an improvement a rifle that could be fired and loaded lying prone. He would have approved of khaki uniforms; after all, he had overseen the introduction and first use of camouflaged troops, the Rifles wearing green as an attempt at concealment. Lest it be doubted that the British army was firmly wedded to the past, after years of trials it introduced into service in 1913 a new-pattern officer's cavalry sabre; the other ranks' new-pattern sabre had come a little earlier. Nobody seems to have taken any notice of the American Civil War and the devastating effects of massed infantry fire and the move to trench warfare. Truly, the armies of Europe went to war unsuspecting of the horrors of industrialized warfare, despite having had plenty of warning.

But can the same charge be levelled at navies? The nineteenth century was a period of rapidly accelerating change in naval warfare, or rather in warships. While there were small-scale actions between single ships, and shore bombardments such as those conducted by the British and French during the Crimean War, there was nothing that equated to a fleet action since the battle of Navarino Bay in 1827, fought between sailing ships using muzzle-loading guns over ranges measured in tens of yards, and the battle of Tsushima in 1905 during the Russo-Japanese war. Yet during that time navies had moved from wooden ships, not far from the type of ship that took Napoleon into captivity, to steam-powered steel warships with turreted breech-loading guns firing over ranges measured in miles. All navies, and particularly the Royal Navy, took every opportunity to learn from experience, their own and that of others. Foreign actions, such as the battle of Hampton Roads in 1862, the first between armoured warships, the CSS *Virginia* and the USS *Monitor*, were carefully studied and drove forward thinking on armour and the mounting of guns in turrets, as did the action between HMS *Shah*, HMS *Amethyst* and the Peruvian vessel *Huáscar* in 1877. *Huáscar* was British-built and carried her guns in a rotating turret, whereas the *Shah* mounted her guns firing through gun-ports in the side in what was then the conventional style. Add to the mix completely new weapons such as torpedoes, submarines, airships, aeroplanes, mines and so forth, and the new technologies were forcing changes in the ways wars would be fought.

It was not only the technology that had evolved. How to fight a war with a major power, when and if it came, was a major concern. Having defeated the United States by naval blockade in the war of 1812, the Royal Navy was also very aware of how much damage had been wrought by the US war on British trade, a *guerre de course* carried out largely by privateers, civilian vessels licensed by their government. World economics had changed. Since the Napoleonic Wars, the countries of the world had become very interdependent; now they had to have each other's produce to live. Thus, while the armies of the world had plenty of experience to draw on and did not evolve their thinking, the navies of the world sucked up every bit of experience they could and entered the war fully aware that this one was going to be very different from anything that had gone before.

The war, when it did come, forced major social changes on Britain. It also led to a major perceptual change in how the country viewed its two armed forces (three by the war's end). Before the war the Royal Navy had an unquestioned position as the 'Senior Service'. It was seen as being the guarantor of the country's safety and of its extensive empire. The army was seen as an imperial gendarmerie, which had not performed very well in the Boer Wars. There were army heroes, Field Marshal 'Bob' Roberts, Field Marshal Kitchener and others, but they were heroes of minor wars that did not significantly impinge on the empire or its integrity. For the army, to make matters worse immediately before the war, there was the taint of mutiny, the Curragh Mutiny concerned with Irish self-rule.

How Britain came to change the very basis of its centuries-old foreign policy is beyond the scope of this book, except to note that Britain moved within a few years from avoiding major land commitments in Europe to fielding by far the biggest army in its history, and the navy was seen as having barely participated in the war.

This book tries to examine the Royal Navy's performance in World War I from a contemporary perspective rather than in retrospect. For example, before the war intelligent and able people (including Winston Churchill) had reasoned that convoying merchant ships was both impractical and unnecessary. The modern perspective is that convoy was the obvious answer to the German campaign of unrestricted warfare, and so it proved; however, it did not appear so at the time. Similarly it is easy to criticize Admiral Jellicoe, as many have done, for his handling of the Grand Fleet at the battle of Jutland, but no one has proposed realistic alternatives to what he did. The book will attempt to show that the Royal Navy's great achievement in World War I was that it actually got most of the answers right, but it was to be a testing four years. It will be shown that while the major fighting was on the Western Front, the Royal Navy effectively strangled Germany economically. Without its contribution the war could not have been won.

This book, which looks at the Royal Navy in World War I, cannot do so from a standing start. It therefore first looks at the incredible changes in technology in the preceding few decades, which made World War I 'truly, a war like no other'[1] at sea as well as on land. Thus it explores the history of the Service before and during the first great conflict of the twentieth century. For the Royal Navy it was in a lot of ways to be a preparation, even training, for what was to be the greatest test it has ever faced, World War II. I hope I have been able to show that the navy that went to war in 1914 brought to it the originality, initiative, flexibility and raw courage that served it well in both conflicts.

CHAPTER 1

Europe on the Brink of War

There are known knowns; there are things we know that we know. There are
known unknowns; that is to say, there are things that we now know we don't
know. But there are also unknown unknowns — there are things we do not
know we don't know.

Donald Rumsfeld, US Secretary of State
12 February 2002

In 1904 Admiral John 'Jacky' Fisher, the First Sea Lord, predicted that war
in Europe would break out after the summer harvest in 1914; he termed it
Armageddon. Historically the end of the harvest had been the traditional
time for nations to go to war, allowing for a three-month campaign and
for everyone to be home for Christmas. He looked forward to it, not with
any pleasure, but as being inevitable.

In 1914 Winston Churchill, the First Lord of the Admiralty, and Admiral
Battenberg, now the First Sea Lord, agreed that to save money, instead of
the annual summer exercises, the Royal Navy would exercise mobilization,
including reservists, and have a Royal Review of the fleet. At that time a
country mobilizing was seen as being a threat to possible enemies, but as
this mobilization was publicly stated in advance in parliament as being an
economic measure, it attracted little public attention at home or abroad,
although the German navy watched closely.

As the summer of 1914 progressed and war seemed imminent, the
Admiralty did not fully disperse the ships it had assembled and manned.
Thus, when war actually broke out the navy was already mobilized. Britain
was felt to have gained an advantage ahead of what was seen to be the
inevitable and immediate clash of the German and British battlefleets.
As the German army marched through Belgium, the British Grand Fleet
waited for a matching advance by the German *Hochseeflotte* (High Seas

Fig. 1.1. A known unknown at the outbreak of war: how much
of a threat was a submarine to a warship? HMS/M *C13* running
submerged and *Magnificent*, a pre-Dreadnought battleship.

Fleet) across the North Sea and probably into the English Channel in an
attempt to disrupt the passage of the British Expeditionary Force to France,
or possibly to cover a landing of the German army in England itself. But
nothing happened; the North Sea stayed empty of the German fleet. The
naval Armageddon was to be delayed.

Armageddon may have been delayed for Fisher, now in retirement,
and most of the navy, but for some there was to be action. On 28 August
a confused and exciting battle took place in the Heligoland Bight. This was
initially between cruisers and destroyers of both sides, but the intervention
of British battlecruisers commanded by Vice Admiral Sir David Beatty led
to a convincing British victory. British submarines present in some force
played little part in the battle, but within three weeks the navy was to receive
a salutary lesson in how much warfare had changed due to submarines.

In the aftermath of the battle of Heligoland Bight, Commodore Reginald
Tyrwhitt's flagship *Arethusa* had to be towed home by an armoured cruiser,
the *Hogue*, after sustaining battle damage. *Hogue* was one of a class of six
ships which were overdue for scrapping. At the outbreak of war, they
were part of the Third Fleet of reserve ships. These ships in peacetime

had only 10 per cent of their ship's company, and only went to sea during mobilization exercises. For the July 1914 mobilization exercise prior to the war they were brought up to full manning with reservists and cadets. From the Royal Naval College at Dartmouth 434 cadets, in reality school-boys, some as young as 15, had been mobilized and sent to sea, seven of them to each of the ships of the 7th Cruiser Squadron (7CS), which was employed in patrolling an area of the southern North Sea known as the Broad Fourteens. No one seemed very sure what they were supposed to be doing there. The best that Admiral Sturdee, the head of the naval staff, could come up with when asked the question by Commodore Keyes was: 'My dear fellow, you don't know your history. We've always maintained a squadron on the Broad Fourteens.' It is difficult to see what value elderly ships crewed by a mixture of elderly reservists and very young semi-trained sailors were supposed to have; the ships were certainly not fit to fight any likely opponent, and were far too slow to run away if they did encounter one. Having had Commodore Keyes describe them to him as the 'live-bait squadron', Churchill (then First Lord of the Admiralty) had agreed with the First Sea Lord, Admiral Battenberg, that they should be withdrawn. Admiral Sturdee was instructed to do so, but, feeling that they provided a useful early warning if the Germans attempted to strike at the vital cross-channel traffic, he persuaded Battenberg to leave them until they could be replaced by new light cruisers which were due from the shipyards in the next few months.[1]

So on 22 September 1914 three ships of the squadron were at sea patrolling the Broad Fourteens: HMS *Hogue*, HMS *Aboukir* and HMS *Cressy*. Because of rough weather their destroyer escorts were detached, and the same rough weather convinced the officers of the three patrolling ships that submarines would be unable to operate. Thus while they maintained a lookout for periscopes, they did not zigzag as an anti-submarine meas-ure, and because their old engines could only manage a maximum of 15 knots, they were steaming slowly in line abreast. They were sighted by *Kapitänleutnant* Otto Weddigen, in command of the German submarine *U-9*, who was able to operate despite the weather. His first torpedo hit *Aboukir*. The explosion was initially thought to be the result of a drifting mine, and so her captain, Captain Drummond, hoisted a mine warning signal. Captain Nicolson of the *Hogue* actually realized that the explosion had been the result of a torpedo. As *Aboukir* sank, despite knowing that there was a submarine in the vicinity, he stopped his ship to rescue survivors. This made the *Hogue* a static target; she was hit by two torpedoes and in her turn sank. Captain Johnson of the *Cressy* sighted *U-9*'s periscope and opened fire on it. Unbelievably, he then stopped his ship to rescue survivors and was torpedoed in his turn. In total 1,459 officers and men were lost, including 13 Dartmouth cadets and midshipmen.

There was a public outcry about the loss of the ships, which worsened when it became public that not only were the cadets young and untrained, they were actually still officially pupils at Dartmouth, which was a fee-paying school; their parents were paying for them to be at war. The loss of life was even worse than it might have been because Admiralty policy was that ships would only carry one lifebelt for every ten members of the ship's company. This episode contributed to the popular image of the war, held by many today, of brave men led by incompetents.

Admiral Sturdee's career survived his misjudgement, and indeed he went on to be the victor of the battle of the Falklands. The navy had learnt a chilling lesson: that submarines were not the toys some had thought them to be. 'Damned un-English weapons' or not, they were to be taken seriously.

The earlier loss on 5 September of the light cruiser HMS *Pathfinder* to *U-21* was followed by another incident on 15 October, in which Weddigen sank another armoured cruiser, HMS *Hawke*, part of the 10th Cruiser Squadron. Now the pendulum swung too far the other way, and for a time almost every wavelet or dolphin was seen as a periscope, a condition that became known in the fleet as 'periscopitis'. This followed on from a report that a U-boat had been spotted inside the Scapa Flow anchorage. Jellicoe asked the Admiralty for permission to move the Grand Fleet to two more easily defended anchorages, Loch Na Keal in the Western Isles and Lough Swilly in Ireland, until Scapa Flow could be made submarine-proof. Additionally, instead of cruising at an economical 8 knots, ships of the Grand Fleet now steamed at 12 knots or faster to make the submarine's task more difficult. Unfortunately, while the Grand Fleet was in the western waters, the brand-new battleship *Audacious* hit a mine and was sunk.

Even Admiral Beatty personally was to fall victim to this scare in January 1915, not far from where the *Hogue* and her sisters were sunk, at the battle of Dogger Bank.

Fig. 1.2. HMS *Aboukir*, one of the 'live-bait squadron', the 7th Cruiser Squadron. With HMS *Hogue* and HMS *Cressy* she was sunk by *U-9*.

Fig. 1.3. A 1912 picture of cadets at the Royal Naval College, Osborne. These boys would have been at Dartmouth when war broke out; all went to sea, some to the 7th Cruiser Squadron.

Fig. 1.4. The German submarine *U-9*.

Fig. 1.5. The brand-new Dreadnought HMS
Audacious sinking after hitting a mine.

The coming of the war

Despite Fisher's forebodings, by the time 1914 had actually come, viewed from England, war in Europe did not seem inevitable. There had been 'alarums and excursions' as Shakespeare would have had it, most recently at Agadir in 1911, when Germany had come close to provoking a war, but the received wisdom was that the world was too economically integrated for any rational person to envisage a war between the major powers.[2] Certainly there were those who wanted war; some French thirsted after revenge for their defeat in the war of 1870 and the return of the provinces of Alsace and Lorraine lost to Germany. The Balkans were a continuing concern; there had been a series of wars between various countries in the region, but these had not involved the major powers and no one thought that they would be drawn into another Balkan War.

However, the German General Staff, obsessed to the point of paranoia with Germany being encircled by enemies, saw a relatively short window of opportunity before they would be overwhelmed, and actively desired and worked for a war. While Bismarck as German Chancellor had fought a series of short, sharp wars, his underlying policy had been first to unite Germany, which he had achieved, and second to isolate France, seen as Germany's most likely enemy, from possible allies diplomatically. Britain herself wished to follow a policy of detachment from the Continent – friendly to all, but allied to none.[3]

Over the 20 years following Bismarck's departure, Germany had from almost nothing built a large and very capable navy, which was probably the only way she could have antagonized Britain. It is difficult over a century later to accept that the build-up of the German navy and the resulting Anglo-German naval rivalry was largely the product of one man's will, that of Kaiser Wilhelm II, grandson of Queen Victoria. He had a love–hate relationship with England and the British royal family and had decided that Prussia, the dominant power in the federal German state, which had historically been a land power, needed a navy, which also served as a means to unify the nation. Aided and inspired by Admiral von Tirpitz, the Kaiser had forced a compliant Reichstag to fund an enormous building programme. Ostensibly this was to support the nascent overseas German empire, their 'place in the sun', but rather than the long-range cruisers such a policy would require, the *Hochseeflotte* was built around shorter-ranged battleships and battlecruisers which were obviously built to fight either France or, far more likely, Britain.

Germany did not need a navy: Britain did. She was a maritime trading nation with an extensive merchant fleet and an enormous empire to protect; she was also, of course, an island. The result was a naval building race between the two. Britain made many efforts to negotiate a slow-down as the money for the ships came at the expense of social programmes to which the Liberal government of Herbert Asquith were committed. The Kaiser, head of an autocratic government, refused any discussions about mutual limitations of naval spending. Thus by the time war broke out, the British Admiralty was, in terms of expenditure, the biggest government department.[4]

Technological change: 'known knowns'

It is difficult a century later to comprehend fully the changes that had affected the navies of the world in the nineteenth century. Naval technology had evolved relatively slowly until the end of the Napoleonic Wars, but then the industrial revolution began to take hold. The most obvious changes were in the use of iron and steel for the structures of ships, and steam power. Initially iron was only used to supplement and/or replace components such as hanging knees in wooden ships, and subsequently as armour on a wooden ship, as in the French ironclad *Gloire*, launched in 1859. HMS *Warrior* in 1860 was constructed of iron, but still had a full sailing rig.

As the reliability and efficiency of steam propulsion improved, HMS *Devastation*, the first major warship reliant totally on steam propulsion without masts and sail-carrying yards, was launched in 1871.

The change from wood to steel and from sail to steam would in themselves have required a complete change in the thought processes of naval officers. Occurring in parallel, developments in gunnery meant that the Nelsonian tactics of laying your ship at half a pistol shot from the enemy and pounding them with muzzle-loading cannon, while now history, were very recent history. Even the famous inconclusive US Civil War battle between the *Virginia* and the *Monitor* in 1862 had been fought at almost point-blank range.

Fig. 1.6. HMS *Warrior*, built in 1860 of iron and with a full sailing rig, which could propel her at 13 knots. Her 5,267-hp steam engine would allow her to steam at 14 knots, but her range was limited because the engine was inefficient, and so the ability to sail remained essential.

Fig. 1.7. HMS *Devastation* in 1871, launched a mere 11 years after HMS *Warrior*, was the first British battleship without sails.

Known unknowns

By 1905 the battle of Tsushima, fought between the Japanese and Russians, was a portent of change. The ships were steaming at about 15 knots, five times the speed at which Nelson had closed the combined fleets at Trafalgar. At Tsushima, firing commenced at ranges initially of about 6,000 yards, or 3 nautical miles. This was the result of major developments in metallurgy and engineering: stronger barrels and breech-loading had been made practical, and, more importantly, there had been significant developments in the chemistry of propellants. Unlike gunpowder, which explodes, cordite burns very rapidly, making for a rapid build-up of pressure inside a gun rather than a near-instantaneous one. To make use of this guns became longer, which made for greater range and greater accuracy. This accuracy, however, was useless unless the spot where a shell would fall could be predicted. Rather than be guessed, the distance to a target – its range – had to be measured, which demanded a very high quality of optical instruments. Then allowances had to be made for obvious variables like the firing and target ships' speed and course, as well as wind, air temperature and direction; shells were now in the air long enough for the rotation of the Earth below them to matter. This process was known as 'fire control'.

Modern perceptions of the resulting naval officer have been shaped by the view that:

> in 1900 the officer corps of the Royal Navy displayed the characteristics of professional inbreeding to the extent of Goyaesque fantasy. Arrogance, punctilious ritual, ignorance of technical progress [...] were added to the unchanged organisation of the eighteenth-century navy to produce a decadence hardly matched in any force of modern time.[5]

This is a travesty. In this instance it was Royal Naval officers who had led in the development of analogue computing devices for gunnery control. Lieutenant John Dumaresq designed an eponymous calculator and Captain Frederic Charles Dreyer designed a fire control table that worked with it. The Admiralty and naval officers had been responsible for many advances in naval technology, from gunnery to torpedoes and engines.

The advent of the torpedo, originally termed the 'locomotive torpedo'

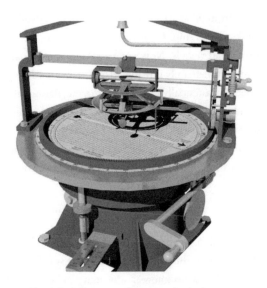

Fig. 1.8. A Dumaresq fire control calculator,
designed by a Royal Navy lieutenant.

Fig. 1.9. The torpedo, another known unknown. Was this
new weapon going to make the battleship redundant?

to distinguish it from what are now known as mines, introduced a weapon that could sink a battleship. Of greater concern was that they could be launched from small vessels. Now any vessel could attack and potentially sink a battleship, hitherto only threatened by another battleship. Those small vessels included the potentially invisible submarine. Because they were small and relatively cheap, they were seen as the weapon of the weaker power, or even by Admiral Lord Charles Beresford as 'playthings' or 'toys'. Nonetheless, they threatened the dominance of the battleship. Another major change was the advent of wireless. Now ships could be directed out of visual range.

1914: Unknown unknowns

World War I was truly a war like no other. The armies embarking on it were at best dimly aware that this was to be a different war from those that had gone before, but the navies knew it was going to be very different. They also knew that they did not know how the various changes were going to affect their war. They did not know if the basic tenets of naval warfare that had held true for centuries were to be modified or changed, and if so, to what extent. For example, it was obvious to most that a close blockade of an enemy coast, such as had defeated Napoleon and the United States of America in the war of 1812, was now impossible, but acceptance of a distant blockade of Germany meant that German ships could range the North Sea before being caught. Was the navy to accept that German ships could bombard British coastal towns? When they established a distant blockade, how would they deal with neutral shipping? The Royal Navy had more submarines than any other navy, but it still had to develop a workable doctrine for their use. At the outbreak of war it was planning to use them in groups commanded from a surface ship. How was a massive surface fleet travelling at 20 knots and able to fight enemies up to 10 miles distant to be commanded, directed and controlled? Even more worrying, had the correct types of ships been developed? There were those arguing for swarms of small ships, so-called 'flotilla defence', rather than vulnerable battleships. How were airships and aircraft to be used? As the fleet dispersed to its war stations, these and many more questions remained unanswered. However, there was, in the opinion of the navy and the wider public, a known known: the high quality of the officers and ratings of the Royal Navy.

Chapter 2

The Royal Navy at the Outbreak of World War I

The earth is full of anger,
The seas are dark with wrath

Rudyard Kipling,
'Hymn before Action'

During the century following World War I, public and academic perceptions of the war have gone through a series of changes. Initially the perception was of a war that had been won on the Western Front, which despite appalling losses was won by good if not great generals such as Field Marshal Haig. The war at sea was largely seen as being a sideshow. With Haig safely dead, the ex-Prime Minister David Lloyd George led the charge, suggesting that it was only civilian intervention (with him in the forefront) that had reined in and directed the stupidly incompetent generals. That view took hold and persisted until very recently. It is now gradually being realized that it was the British army that played a major role in the military defeat of the German army, and that the generals, in the circumstances, actually did quite a good job.

Perceptions of the Royal Navy's role in the war have been slower to change, but even now the naval war is still regarded as peripheral to the 'main war'. Those who have heard of the Battles of the Somme and Passchendaele are often at best barely aware of the Battle of Jutland, let alone the major role that the navy played in the defeat of the Central Powers. What has not helped is the widespread belief that the navy was living on past glories, the 'long calm lee of Trafalgar', and had resisted change before being dragged, almost kicking and screaming like a wilful child, by Admiral Fisher into the twentieth century. What is forgotten is that the wooden sailing navy of 1805 (the year of Trafalgar), which was developmentally not far removed from that of the Armada in 1588, had, in a century, changed totally, and with these changes, so too had the demands placed on its men. To understand the Royal Navy in World War I, it is necessary to start by looking at its people; despite the old saw that the army equips its men, whereas the navy mans its equipment, the people were absolutely central to how the navy worked and fought.[1]

The men

The army and the Royal Navy had radically different approaches to the way they recruited, trained and retained their manpower, both commissioned officers and men. In the army there were officers and other ranks, and movement up the hierarchy was by promotion. It was possible for an 'other rank' to become an officer and receive the Sovereign's commission; indeed Field Marshal Robertson, who was to head the army during World War I, had started his career as a private soldier.

The Royal Navy was divided into officers and ratings: the former moved upward by promotion; the latter, described as 'the lower deck', by advancement or being 'rated up'. The greatest difference from the army was that the divide between officer and rating was nearly complete. There were 'warrant officers' in the navy who had originally been ratings and were termed subordinate officers, and who wore narrow gold-lace rank insignia and a smaller version of the officer's cap badge. Their authority derived from an Admiralty warrant, not the Sovereign's commission. Within the executive branch, warrant officers were narrow specialists, for example gunners, signal boatswains and so on. It was theoretically possible for a warrant officer to become a 'commissioned warrant officer' and from there to progress to a lieutenant. In reality, however, between 1818 and 1913 it was virtually impossible for a rating to become an officer.

Within the rating structure there were major distinctions; the biggest was between the seamen and the stokers. Stokers not only stoked the boilers, they also provided men for the engine rooms. In the popular perception a stoker was the man who carried coal from the stokehold where the coal was stored and shovelled it in to the boilers. In reality, while he did do that, he was far more than a labourer: he was a skilled man. A ship such as the battlecruiser HMS *Lion* had 42 boilers. Unlike a merchant ship, which would get up steam over many hours before sailing and then travel at a steady speed from port to port, warships had to be able to get up steam very rapidly and vary the amount of steam as they changed speed to allow them to manoeuvre. Merchant boilers could be 'banked': that is, piled up with coal which steadily burnt away. Warships needed the coal to be spread carefully across the bed of the boiler, and the much higher steam pressure (in *Lion* at 235 pounds per square inch) had to be monitored carefully. The stoker was a skilled man, and was paid accordingly, significantly more than a seaman.[2]

The Royal Navy was at its full manpower strength when the war broke out, having already mobilized. It was made up of regular sailors who were very largely long-service; that is, they had signed on to serve for 12 years with the option of extending to complete 22 years service, upon which they would leave the navy with a pension. When the navy mobilized for war they were augmented by men of the Royal Fleet Reserve (RFR). These men were former ratings who had completed an engagement and returned to civilian life, but were now recalled. Another group was the officers and ratings of the Royal Naval Reserve (RNR), who were men of the merchant service who had agreed to Royal Naval service in the event of war.

Unlike the army, which expanded enormously during the war, the navy did not. By the end of 1917 nearly three-quarters of the seamen and 78 per cent of the stokers were regular or ex-regular. Thus the lower deck, despite three years of war, was still largely composed of regular sailors. While there was a marked expansion of the Royal Naval Volunteer Reserve (RNVR), most of them went into the Royal Naval Division (RND) and other areas seen as being peripheral to the navy proper. Particularly later in the war, many were what would now be called communications specialists, recruited specially to run the 'Y' and the 'B' wireless monitoring services.[3]

The officers

In Nelson's day, an officer could be, and was expected to be, a master of every area of his profession. By 1914 the profession of naval officer had become so diverse that officers had to specialize and senior officers had to be advised; they could no longer rely on their own knowledge and

experience. Admiral Fisher had considered how to manage this, and introduced a structure that had officers specialize after becoming lieutenants.
They would then spend some years undertaking specialist duties, including
advising their seniors as staff officers, training others and serving in the
Admiralty, which might involve the procurement of ships and equipment.
Then, on reaching higher rank, either as a commander or captain, they
would return to be a generalist, as a leader and manager.

The navy had changed the education and training of its officers from
being an apprenticeship at sea to an education ashore, ultimately at the
Britannia Royal Naval College at Dartmouth, which opened in 1905. This
was supplemented by the Royal Naval College, Osborne, on the Isle of
Wight, which took boys for their first two years. While these were naval
colleges commanded by naval officers, and although the cadets wore naval
uniform, in reality together they constituted nothing more than a public
school with a naval flavour; cadets actually received very little naval training
until they went to sea, but their education was technically biased – after
all, the navy needed technologically capable officers. Voices were raised by
some, such as the future Admiral Richmond, that cadets were not being
given a broad enough education, particularly in naval history, which would
teach them to think.

However, one thing had not changed: the navy remained firmly wedded
to the idea that it had to catch its future officers young. It took boys at
the age of 12, educated them and then trained them to be officers. Thus
Dartmouth was the only source of officers. It took about ten years from a
boy entering the system until he became useful to the navy as a lieutenant
with a bridge-watchkeeping certificate, which meant that he could be an
officer of the watch at sea. When in the last years of the nineteenth century
the navy started to expand rapidly in response to the German threat, there
were nowhere near enough officers. More had to be found, more quickly
than the Dartmouth system could produce them.

As well as using merchant service officers of the Royal Naval Reserve
(RNR), the Admiralty adopted two additional measures to deal with the
shortfall. Firstly, in 1912 it set up the 'Mates Scheme', which allowed ratings
to become officers. In the socially rigid aftermath of the Edwardian era,
this was a groundbreaking change. While political pressure undoubtedly
played a significant part in this, the changes were actually put forward by

the then First Sea Lord, Admiral Lord Battenberg. Potential mates could be selected for promotion from warrant officers, petty officers, leading ratings and those 'who have fully qualified for advancement to Leading Rating'. This was based on recommendation by the '[c]aptain of a seagoing ship'. They would then be successively considered by a committee of officers, convened by a squadron flag officer, and the Admiralty. Once selected, they would be promoted to acting mate, and go on various courses at the Royal Naval College at Greenwich. On satisfactory completion of these courses they would be confirmed as mates before going to sea. Marks gained on these courses would shorten the time until promotion to lieutenant. For example, if he gained ten marks, he would be promoted after two years and two months; four or less, three years and two months. Ultimately, however, promotion to lieutenant was subject to his gaining a certificate of professional proficiency and a certificate from his captain stating: 'He is recommended as in all respects fit for promotion to the rank of Lieutenant.'[4]

Most of the 371 of those promoted to be mates up until 1932, when the scheme ended, were actually commissioned during World War I. A 1914 navy list shows 45 who were promoted in 1913 alone. Promotion to mate continued throughout the war; for example, 32 were promoted in July 1918. Of the 1913 seniority mates, all but two were lieutenants by the end of the war; what happened to those two is unknown. Between them they had been awarded one Albert Medal (Lieutenant Rutland) and six Distinguished Service Crosses (DSCs) the latter for bravery in the face of the enemy, which shows not only that their service had been satisfactory, but more crucially that they had been on active service.[5]

The other new source of officers was the 'Special Entry' scheme, also known widely as the 'Churchill Scheme'. The Admiralty decided to enter boys directly from public schools at the age of 18. The first batch of 30 cadets joined in September 1913, and initially their training was undertaken afloat in a cruiser, HMS *Highflyer*. On the outbreak of World War I their training was moved ashore to the Royal Naval College at Keyham, Plymouth, where it remained throughout the war. After the war there were attempts to abolish the scheme, but it survived and ultimately it was the 12-year-old Dartmouth entry that ceased.[6]

The commanders

The rapid expansion of the navy meant that not only was it short of officers, but it was also short of senior officers. To be the commanding officer of a large ship required a significant amount of experience and practice, so the navy had only one source for its senior officers: its junior officers. This was to be the navy's greatest weakness in World War I. There were officers reaching higher ranks than might reasonably have been expected, or, more worryingly, than should have done. Furthermore, there was a 'lack of discrimination in the training and promotion process', such that 'several classes' from the naval college at *Britannia* achieved 100 per cent promotion to the rank of captain. In contrast, between the wars only about 25 per cent of executive-branch officers were eventually promoted to captain. This meant that many officers who were realistically unfit for the job were promoted.[7]

That assessment is a little unfair, however: the navy had to have senior officers to fill command positions, and it had a limited pool to draw on. If it needed a certain number of captains, it had to promote that number of commanders. This contrasted with the army; becoming a senior infantry officer could be done with on-the-job training, and many did in World War I. Brigadier General Asquith started the war as a junior (naval) officer, plucked straight from civilian life. It would be inconceivable that a civilian could enter the navy and, even in wartime, acquire the expertise and experience to be the naval equivalent of a brigadier general, a commodore.[8]

The navy's mechanisms for promoting officers were actually quite good. It was in no one's interest to promote a dud, even if he was the son of a friend. A century earlier Admiral Collingwood, Nelson's deputy at the battle of Trafalgar, said of a young officer:

> [I]t is a pity [his mother] had not put him apprentice to [...] the apothecary [...] His gravity would have established his reputation [...] and if he did poison an old woman now and then, better to do that than drown an entire ship's company at a dash by running on the rocks.[9]

While favouritism did not play very much of a part in promotion, the bar was being set lower. There were some poor, if not downright incompetent,

Fig. 2.1. HMS *King Edward VII*, the name ship of the last class of British
pre-Dreadnoughts, totalling eight ships. During World War I they
served together as the 3rd Battle Squadron, known as the 'Wobbly
Eight' because of their poor steering characteristics.

Fig. 2.2. HMS *Dreadnought*, commissioned in 1906, made every
other battleship in the world obsolete. In fact, for many years the
term 'battleship' referred to ships before *Dreadnought*; those coming
after were generically 'Dreadnoughts' or 'super-Dreadnoughts'.

senior officers. Some were very brave, but far from bright – a dangerous combination. This was compounded by the navy's intentionally leaving the further education and training of its officers in their own hands; once they had a watchkeeping certificate, they did not have to attend any courses or undertake any training unless they volunteered to do so. The result was some senior officers with a very narrow breadth of knowledge; if Admiral Beatty had had first-hand knowledge of the capabilities of a submarine, one wonders whether he would have acted as he did at the battle of Dogger Bank when he thought he saw a periscope.

Thus the navy entered the war with some weak, if not incompetent, senior officers. The indications are that the Admiralty knew it, and put them in jobs that had to be done, but where they could do little damage. As will be seen, this was not always possible, and with changes of circumstances an officer thought to have been 'parked' would be thrust forward. Another unknown.

The ships and their weapons

HMS *Dreadnought* gave her name to a completely new type of battleship. Until she was commissioned in 1906, battleships carried a variety of calibres of guns. The last British pre-Dreadnought class, the *King Edward VII*, carried four 12-inch, four 9.2-inch, ten 6-inch, fourteen 12pdr and fourteen 3pdr. She had triple-expansion engines, enormous steam piston engines, which drove her 16,350 tons at 18 knots. *Dreadnought* carried ten 12-inch guns and twenty-seven 12pdr. Almost as big a revolution as her gun armament was that she was propelled by steam turbine engines, which pushed her 17,900 tons at 21 knots, and unlike earlier ships she could sustain this speed for extended periods. It has been said many times that she made every other battleship in the world obsolete; she was faster and could outgun any of them.

Battlecruisers have been the subject of much debate ever since their inception. What is largely forgotten is that they were intended as cruisers, built for scouting, and indeed some authors have argued that they were intended to hunt down commerce raiders. For that purpose they were fast and capable of long-range steaming. As all ships are a compromise, in exchange for the requisite speed battlecruisers sacrificed protection and were relatively lightly armoured. The latest battlecruisers were faster than battleships, and like them carried the new 13.5-inch gun. This gun was more accurate than the earlier 12-inch 50-calibre barrels (their length was 50 calibres, i.e. 50 times 12 inches, or 50 feet), the shells of which tended to tumble in flight. HMS *Lion* was a full 120 feet longer than the contemporary *Orion*-class battleship and made 27 knots on trials, a full 7 knots faster than the *Orion*. They were armed like a battleship, looked like a battleship and ended up being used like a battleship. That said, the defects resulting in the losses of three British battlecruisers at the battle of Jutland may well not have been due to design inadequacies, but rather to errors in turret drill.

The Grand Fleet

The build-up of the German Fleet had had two effects. Firstly and most obviously, there was a British warship-building programme aimed at maintaining British numerical superiority. Secondly, with the passing of years the Royal Navy was increasingly concentrated in home waters. Even the Mediterranean Fleet was progressively cut back. Not only were ships brought back into home waters, but the fleets in home waters were also concentrated.

The organization immediately prior to the war was that there were three fleets. The First Fleet was made up of the newest and most capable ships. They were at all times fully manned, and in modern parlance 'worked up'. The Second Fleet was partially manned to 60 per cent and made up of older ships, for example pre-Dreadnought battleships. The Third Fleet comprised older ships not far from the scrap yard, which only had what was termed a 'nucleus crew'. Together they constituted the Home Fleets, which, when they came together, became the Grand Fleet.

Admiral 'Jacky' Fisher

John Arbuthnot Fisher, later Admiral of the Fleet, Lord Fisher of Kilverstone, is now looked back on as one of the greatest administrators and innovators the navy has ever had. His contemporaries loved him and loathed him in roughly equal measure. He had boundless energy, often dancing at balls until the early hours and otherwise working until late into the night.

Fig. 2.3. John 'Jacky' Fisher, shown here as a vice admiral,
was an undoubted naval genius, loved and loathed by
his contemporaries in roughly equal measure.

He was ruthless with those who crossed him. A master of the pungent phrase, he was not above judicious leaks to the press to get his way. He was an innovator of the first order. While at the Admiralty as Second Sea Lord, responsible for personnel, he introduced the major changes in officer career structure described earlier in this book. In deference to the then otherwise unremembered First Lord, these are known as the Fisher–Selborne reforms.

It was as First Sea Lord from 1904 to 1910 that he completely reformed the material side of the navy and its war organization. Recognizing that Germany was the inevitable maritime enemy, he brought overseas squadrons home and scrapped ships that were too weak to fight and too slow to run away. Inevitably he was opposed from within the service and without, and his long-drawn battle with Admiral Lord Charles Beresford was a *cause célèbre* of its day. He is remembered for his greatest innovation, the all-big-gun battleship *Dreadnought*, and his particular favourite, the battlecruiser. Also less remembered is that it was he who recognized the future impact of the submarine and of torpedo-armed smaller ships on naval warfare.[10]

Within the Admiralty, Fisher reorganized the way the navy would fight wars. He has erroneously been much criticized for failing to develop a naval war staff. In fact there had been a central staff with duties beyond the merely clerical in existence from the nineteenth century. The Foreign Intelligence Committee was formed in 1882 for trade defence; this became the Naval Intelligence Department in 1887, with a wider remit. Then it had two branches: one its intelligence function, the other concerned with war mobilization. By 1902 this had become the Naval Intelligence Division, but it had actually established outstations ashore by the turn of the century, as well as two further branches, one to study defence and strategy and another to look at trade (initially from the defensive aspect), but it would come to look at organizing naval economic warfare. Interestingly, Admiral Battenberg is quoted as writing in 1902 to Fisher: 'The N[aval] I[ntelligence] D[ivision] as now constituted is modelled exactly on Moltke's great creation [i.e. the German General Staff]: its three main Divisions have their exact counterpart at Berlin.'[11]

The naval staff before and during World War I has not been treated well by historians, although recently that has shown signs of change. Admiral

Richmond led the charge. He was an outstanding intellect, and after retirement from the Royal Navy became successively a professor of History and Master of Downing College, Cambridge. Before the war, he was a founder of *The Naval Review*, an in-house journal that has recently celebrated its centenary. He was concerned to the point of obsession that naval officers were taught too much of technical subjects and not enough of history. This view has been echoed and expanded on by others, expressing the view that staff officers were often less than able:

> In the Navy the best officers were chosen for sea command, which was the only route to flag rank. The middle ranks of the Naval Staff tended to get whoever was left after the Fleet had made its choice: at best officers of ability whose health or temperament unfitted them for command afloat; at worst the unemployed or unemployable.[12]

This interpretation of the make-up of the naval staff does not bear examination; indeed it is fair to say that in reality the naval staff made 'substantial contributions [...] to victory in 1918'.[13]

Communications

The changes in the naval staff went hand in hand with development in communications. First the telegraph had made for reliable communications between fixed points on land and between continents via undersea cables. Indeed it had been suggested in the latter half of the nineteenth century, probably flippantly, that a deploying fleet would take with it a cable-laying ship so that it could remain in telegraphic contact with England. Britain was exceptionally, indeed exclusively, well placed when it came to cable traffic. Sir John Pender, originally a Manchester textile manufacturer, had moved into cable manufacture and by a series of mergers built the Telegraph Maintenance and Construction Company. It had a virtual world monopoly of the manufacture and laying of undersea cables and dominated the industry well into the twentieth century, not least because Britain controlled the world supply of gutta-percha, essential for the waterproofing of undersea cables. Nobody laid an undersea cable of any importance without British involvement, and Britain was a nexus for many cable lines; even some

German cables to the United States went through Britain. Almost all the cables that transited the Mediterranean went through Malta, then a key part of the British Empire. To give an example, in order to avoid communicating through Germany to their ally Russia, the French sent their cable traffic via Malta. When Admiral Fisher was commander in chief, Mediterranean, based in Malta, he had a private arrangement with the cable companies to receive copies of all diplomatic traffic. He set up a code-breaking cell to read them, and boasted that he read some traffic before the intended recipient. To make use of the information, he established a war room based on a plotting table in Admiralty House on Malta. By 1902 the Director of Naval Intelligence could locate any named French ship, at sea or in harbour, within 15 minutes. Fisher took this organization with him to the Admiralty when he became First Sea Lord, and in 1905 a war room was established within the Admiralty building. This contained an enormous wallchart which tracked the movements of every warship (British and foreign) and significant merchant vessels. In peacetime it was updated every eight hours, but as war loomed this became more frequent.[14]

Wireless had been invented around the turn of the century, and all British warships were equipped with it. The Royal Navy had led the way on the introduction of wireless at sea; indeed Captain Henry Jackson had worked together and in friendly rivalry with the Italian pioneer of wireless, Guglielmo Marconi, who carried out most of his early experiments in England. By the outbreak of war, the navy was well aware of the capabilities and weaknesses of wireless. It was slow. In 1911 Home Fleet tests had shown that a coded 100-word message took ten minutes to transmit (only five if not coded), but 35 minutes to code and 25 to decode. By 1914 the Royal Navy wanted a coding machine, but tests that year did not produce a satisfactory one. At the outbreak of war it was impossible for anything other than a short message (then still called a signal) to be passed from an admiral to his subordinates in another ship to be used tactically, that is, in a battle. Signals were kept as short as practicable; verbosity was discouraged. However, the technology was rapidly improving.[15]

Jellicoe, the commander in chief of the British Grand Fleet from the outbreak of war, prepared a document designed to deal with most eventualities, the 'Grand Fleet Battle Orders'. While regularly updated and amended, they were inevitably prescriptive. Hence he included: '[I]t may be impossible to get orders through quickly, senior officers must not wait for them, but take the initiative.'[16]

The Admiralty was also very aware that it was possible to listen to others' wireless traffic. To do so, before the war the Admiralty developed a network of powerful wireless listening stations linked by telegraph to the Admiralty. There were two separate listening chains: the 'B' service provided directional information, vital to locate ships, merchant or naval, and the 'Y' service recorded messages for decoding. The Russian capture of code-books from the SMS *Magdeburg*, which were passed to Britain early in the war, was the first such instance that made German wireless signal traffic a fertile source of information, and the Admiralty rapidly integrated the intelligence derived into its war-fighting organization based in the war room. At the outbreak of war, the British cut as many German cables as possible to force the Germans to use wireless to communicate, the easier to be read. Once the importance of the code-breaking organization was recognized, it was moved to Room 40 in the Admiralty, literally next door to the war room itself. Room 40 looked at all aspects of German electronic communications, not just wireless.[17]

At the same time information flowed back to the Admiralty from its own ships and officers, as well as from diplomatic posts overseas. While the terms were not yet in use, the information received would be assessed, assigned relative weight and result in what was to be called 'intelligence'.

Having established the war room, Fisher mobilized it for the first time in 1905, when the Russians deployed their Baltic Fleet to the Far East, where it was ultimately to be destroyed by the Japanese at the battle of Tsushima. The Admiralty war room tracked the Russian fleet, correctly assessing before any other authority that it would go via Singapore rather than Batavia. After the battle the Admiralty correctly estimated the position of the Japanese fleet, which differed from what they themselves had signalled! The war room was used during exercises; it ran an hourly updated plot during an exercise in the North Sea as early as 1908. In 1910 the then First Sea Lord Admiral Wilson used it to control a division of battleships operating off the coast of Spain. This was the first time anything like it had been attempted, and it is hardly surprising that it encountered problems. During exercises before the war and throughout it, unit and ship command-ers assumed that the Admiralty had a better picture of what was happening

Fig. 2.4. The fog of war. HMS *Malaya* firing a broadside; that is, all of
her big guns together. Large guns made a lot of smoke. Combined
with funnel smoke, gun smoke would markedly reduce visibility
during a battle, making flags and flashing-light signals hard to read.
Possibly the extent of this was not appreciated before the war.

than they did – a compliment to the organization, but forgetting it was
they themselves who provided the information from which the picture
was put together.[18]

Changes in technology had caused events in naval warfare to evolve
much faster than in the army. Nelson had had some hours from sighting the
French and Spanish fleets at Trafalgar before his ships came into action. As
will be seen, at the battle of Jutland Admiral Jellicoe had to decide what
formation to order his ships into in a matter of minutes. Furthermore,
because they were good at listening to other people's wireless and making
use of it, the navy was also majorly concerned that other people might listen
to theirs. Add to that the problems of intentional jamming and simultaneous
broadcasts, which would do the same, the Royal Navy maintained the ability to communicate by flags and light. This has led some writers to assume
that the Royal Navy avoided the use of 'W/T' (wireless telegraphy) to the
point of preferring flags:

Fig. 2.5. The wireless experimental ship HMS *Vindictive*
with her extremely tall radio masts.

This apparent British backwardness was probably associated with awareness
of how radio intercepts could be used [...] there were also technical problems. Sets could not be tuned precisely, so radio operators found it difficult
to distinguish signals in heavy traffic.

This, combined with the lack of speed and uncertainty of messages being
received, meant that the navy still used flag signals in action. Orders were
passed using a system just like Nelson's. However, even flags 20 feet square
are difficult to see at a distance in good weather, let alone when there is
mist (common in the North Sea) and funnel and gun smoke. There was
an alternative: flashing light using Morse code. This was to a large extent
directional, for it could only be sent to one recipient at a time; multidirec-
tional masthead flashing lights were just not powerful enough.

Study of Admiral Jellicoe's 'Grand Fleet Battle Orders' shows that flags
were regarded as a subsidiary, although important, method of communication. Had it been the primary method, it would probably have been updated
regularly; the extant signal book at the outbreak of war was that of 1897,

albeit with later amendments. Thus while the Germans almost exclusively used wireless tactically, in 1914 the British still used flags and light. Because they were aware of how it could be used against them, they were notable for their good W/T discipline, which steadily improved further.

The changes in communications meant that by 1914 wireless could communicate from shore to ship fairly reliably. This led to further organizational changes. Over the twenty years before World War I, there had been a revolution in the way that navies were commanded and controlled. Until the changes, flag officers in command of fleets as well as captains in command of ships operating independently, so-called 'private ships', had a large degree of autonomy from the Admiralty. Indeed they had to have this autonomy; there was no way they could be controlled minute by minute or even week by week. They would be given general instructions, updated and amended by handwritten letter. Actually the Admiralty was no stranger to more distant communication. Even during the Napoleonic Wars it had a system of semaphore stations that could communicate extremely rapidly with the major ports (if the visibility was good), so that being able to communicate by cable and then by wireless was to some extent a step change, not something totally new. What was more, it was adaptive to the restrictions imposed by the new technology. When Admiral Sturdee was dispatched to the South Atlantic in search of Admiral Graf von Spee, it was realized that he would be out of wireless range of the Admiralty. The navy dispatched its wireless experimental ship HMS *Vindictive* – 'at a time of very tall wireless masts, hers were tremendous' – to sit halfway down the Atlantic, acting as a relay ship.[19]

What did happen as a result of the development of a central naval staff with the capability to command and control ships at sea was an alteration in the command structure. Previously, area flag officers controlled everything that happened in their area; thus the commander in chief, Portsmouth, controlled the ships operating in his area. Flag officer, Malta, was also the commander in chief of the Mediterranean Fleet, and so on. Now there was recognition that the Admiralty would take more direct control. The Grand Fleet, whose formation is to be discussed, had a commander in chief who was responsible to the Admiralty. Other commands such as the Harwich Force eventually came under the Admiralty, not the senior naval officer at Harwich. This system was actually quite flexible and, as will be seen, was amended quite rapidly as and when the situation required it.

Central direction of the armed forces

While there was cooperation at lower levels between the two services, there had been little at the upper levels. This was probably because it had not been deemed necessary, as each service operated in its own domain. In fact Fisher had described the army as a projectile fired by the navy. That the navy would cooperate once ashore was also well established. What was not seen as necessary – and did not happen – was cooperation at the military's top strategic level. That is not to say there was not planning at the top level where the politicians became involved; what would now be termed the grand strategic level. The defence of the British Empire was

Fig. 2.6. Winston Churchill in 1915 as First Lord of the Admiralty.

overseen by the Committee on Imperial Defence (CID) established in 1902. This body was chaired by the Prime Minister or the most senior of the two service chiefs, either the Chief of the Imperial General Staff (CIGS) or the First Sea Lord.

The committee became increasingly important. Its secretary at the outbreak of war was an ex-Royal Marine artillery officer, Maurice Hankey. He was to be an increasingly important and powerful figure, eventually becoming Cabinet Secretary and Secretary to the Privy Council, the three most important committees in the Empire. Starting in 1909, he was responsible for creating and maintaining the 'war book'. Each government department had a chapter in the war book, which contained a summary of all the actions to be taken by the department on the outbreak of war, together with a synopsis of the corresponding and simultaneous actions to be taken by other government departments.[20] The book was continuously updated under the oversight of the CID and allowed Britain and its empire to transition smoothly to war in 1914.

Agadir and the arrival of Churchill

The 'Agadir incident' was named for a North African town where the German navy staged a provocative incident that nearly provoked war and caused Britain to examine its war preparations. Historically, Britain had avoided getting involved in Europe unless it had to; it had certainly avoided Continental alliances. However, it had found itself increasingly tied, although not by treaty, to France. The ties had included talks between the two army staffs, without commitments. After Agadir, Herbert Asquith, the Prime Minister, called a meeting of the CID so that he could be briefed by the two services about their war plans. General Wilson, Director of Military Operations, gave a polished performance indicating that the army would move six divisions to France. After lunch it became apparent that there was a profound disagreement between the navy and the army as to how the war, if it came, should be fought. The navy envisaged fighting an economic war against Germany, which had been approved by the Cabinet as early as 1908. Admiral Wilson, the First Sea Lord, did not by all accounts give a good performance when putting forward the navy's case for avoiding deploying an army to the Continent. It is almost as though it was the first time he had realized that the army's intention was to fight alongside the French on land. It does appear, however, that no formal decision was actually taken by the government, because when in 1914 war became imminent, as Churchill records, at least three-quarters of the Cabinet 'were determined not to be drawn into a European quarrel'.[21]

The upshot of the meeting was that Churchill was appointed First Lord of the Admiralty, its political head. When appointed, he was an ambitious 36-year-old, highly intelligent, extremely articulate and very convincing in argument; there is no doubt, for instance, that he would have made an outstanding trial lawyer. He brought to the post the experience of extensive military service at a junior level, but he had had no naval or maritime experience. Prior to coming to the Admiralty, Churchill had received a selective naval education at the hands of Admiral Fisher, who had been First Sea Lord, the uniformed head of the navy. They had met socially in the first instance, and Fisher, spotting a rising talent, had sought to convert Churchill to his cause(s).

Very nearly Churchill's first act on coming into office was to sack Admiral Wilson as First Sea Lord and replace him with Admiral Bridgeman. He then replaced almost the entire Admiralty board. As a worrying portent of his future management style at the Admiralty and afterwards, he appointed Rear Admiral David Beatty as his naval secretary, or in modern terminology, his executive assistant. Beatty was extremely personable, an outstandingly charismatic leader, vain to a fault, married to a very wealthy wife, and shared Churchill's passion for polo. He had recently turned down an offered appointment at sea, and in the normal course of events he would not have been offered further employment had Churchill not picked him out.

Then he 'never ceased to labour at the formation of a true General Staff for the Navy'. At the end of it, in January 1912 Churchill produced a memorandum 'on the reorganisation of the naval Staff and the establishment of Staff Training'. It is obvious that Churchill had been climbing a very steep learning curve, first having learnt the lesson that 'the sea is different'. In the memorandum, he effectively accepted that there was already a naval staff in existence, as has been described, which functioned to advise the First Sea Lord. Essentially what happened was that a large section of the Naval Intelligence Division was renamed the Naval War Staff. However,

Churchill was a politician. He was not the first and will not be the last to make a molehill into a mountain and triumphantly climb it.[22]

War

As already mentioned, the fleet (meaning the totality of the Royal Navy) had mobilized in July. This mobilization was supposed to have finished with a royal review of the fleet, following which the ships were to have dispersed to their peacetime employment. However, Battenberg kept the First Fleet concentrated at Portland, while Churchill reversed the demobilization of the older ships. When war broke out the fleet was already at wartime readiness.

Before Britain actually declared war on Germany, the Foreign Secretary Sir Edward Grey announced in the House of Commons that the German fleet would not be permitted to enter the English Channel. The British government had to do this as a minimum; they had earlier agreed with the French that in the event of war the Royal Navy would be responsible for the Channel, and accordingly the French had placed most of their ships in the Mediterranean. After much heart-searching in the Cabinet and the resignation of two of its members an ultimatum was issued to Germany,

Fig. 2.7. The royal review of the fleet, 20 July 1914. HMS *Iron Duke* and HMS *Marlborough* are both wearing the flags of a full admiral.

demanding the withdrawal of their troops from Belgium. This was to expire at midnight on 4 August. At 17:50 the Admiralty sent a signal to all ships stating: 'The war telegram will be issued at midnight authorising you to commence hostilities against Germany, but in view of our ultimatum, they may decide to open fire at any moment.'[23]

Having already mobilized, the Grand Fleet, made up of the fully manned First Fleet and the ships at lower readiness, deployed to its war station at Scapa Flow. There was something of a stand-off; the British expected the Germans to come out, the Germans that the British would institute a close blockade. To warn against it and to provide an early warning they deployed their submarines into the North Sea. But neither did what their enemy expected.

The Grand Fleet was commanded by Admiral Sir George Callaghan. During his period in command he had done well to improve the fleet's readiness for war, so much so that in December 1913 his command was extended for a further year. He knew that Vice Admiral Jellicoe would relieve him in the ordinary course of events in December 1914. Churchill, however, felt that an immediate change was necessary, as he alleged that at the age of 62 Callaghan was too old for what would be an arduous and demanding job. He told Callaghan that he was sending Jellicoe as his deputy, and sent Jellicoe with sealed orders to take over from Callaghan. Properly, Churchill did not have the authority to make this change by himself. All flag-officer appointments were (and still are) required to have the approval of the Sovereign, which Churchill sought during Jellicoe's journey.

John Jellicoe was then 54. He was a gunnery specialist and was widely regarded, including by Fisher, as the best naval officer of his generation. Callaghan was a close friend, and Jellicoe did his best by telegram at various stops en route to Scapa Flow to dissuade Churchill. Fortunately Callaghan made the handover as easy as possible in the circumstances and quietly left the fleet that he had prepared for the coming war as it was leaving harbour to go to sea under its new commander in chief.

Churchill had likely made the right decision. Jellicoe was probably the better admiral, as the annual fleet manoeuvres in July 1913 had shown. Whether Churchill had made a good decision when choosing the commander of the Battlecruiser Squadron is more debatable. It will be remembered that on becoming First Lord of the Admiralty in 1911 Churchill had chosen as his naval assistant Rear Admiral David Beatty. In 1913 Beatty became the commander of the Battlecruiser Squadron, widely regarded as a plum appointment. He got the appointment over the heads of more senior and, many felt, better-qualified officers, but the battlecruisers – the fast, heavily armed scouting arm of the Grand Fleet – might have been made for a man of Beatty's temperament. He was to fly his flag aboard *Lion*, the first of the 'splendid cats': *Lion*, *Queen Mary* and *Princess Royal*, and *Tiger* when she joined the fleet. During most of the war they were based on the Firth of Forth in the new dockyard complex at Rosyth. They were still part of the Grand Fleet. Being based further south and thus closer to German waters was obviously an advantage, but it was also a disadvantage. Grand Fleet ships had ready access to areas where they could practise firing their main armament. For Beatty's ships this was not so easy, and many have

Fig. 2.8. HMS *Lion*, Beatty's flagship while he commanded the battlecruisers. The two turrets visible at the forward end of the ship are 'A' and 'B' respectively. 'Q' turret, which was to be hit at the battle of Jutland, is between the funnels. A second turret amidships would be 'P'; after turrets would be 'X' and 'Y' (*Lion* only had one). All ships used this nomenclature except HMS *Agincourt*, whose seven turrets were called for the days of the week!

ascribed the battlecruisers' poor shooting in comparison with that of the rest of the Grand Fleet to this cause. As will be seen, the attempt to make up for this deficiency by more rapid fire was to have severe consequences.

The Battlecruiser Squadron's role was to find the enemy and guide the Grand Fleet to them. Communications were thus of the highest importance. It is curious therefore that as his flag lieutenant responsible for his communications Beatty chose Lieutenant Ralph Seymour, who was promoted in the ordinary course of events to lieutenant commander in December 1914. A very personable officer, Seymour was very well fitted to undertake the secondary duties of his post, managing the demanding protocol aspects of his admiral's duties, although he was not a signals specialist. Regrettably over the next few years this deficiency was to become all too apparent.

Another significant part of the British fleet, but under Admiralty command through the admiral of patrols and thus not under the Grand Fleet, was the Harwich Force. This was commanded by one of the best officers the Royal Navy was to produce during World War I, Commodore Reginald Tyrwhitt. He was ultimately to be an admiral of the fleet. At that time commodore was not a rank, it was an appointment; thus Tyrwhitt was still a substantive captain. To give a measure of the man, he was the only captain in the twentieth century to be knighted while serving at sea. His command ultimately included 15 light cruisers, and three flotillas of destroyers (the 1st, 2nd and 10th), each of 20 ships. In addition he had attached to his command four seaplane carriers. This force was steadily augmented during the war and was to be of major importance, particularly in the opening phases.[24]

Commodore Roger Keyes also worked out of Harwich. He had been inspector of submarines and had been responsible for a lot of the development of the 'E' or 'Overseas' (i.e. long-range) submarines, despite himself not being a submariner. Keyes was a charismatic and inspiring leader, although Admiral Richmond felt that he was impulsive to a fault.[25]

The Admiralty had initially set up a command arrangement at Harwich that while the Harwich Force came directly under the Admiralty, the 8th Submarine Flotilla based there, comprised initially of ten submarines, two depot ships and two destroyers, came under the command of the Grand Fleet. This reflected the thinking that the submarines would scout in advance of the fleet, controlled from one of the destroyers. To modern thinking, tying a submarine to a surface ship loses it its great advantage of manoeuvrability and invisibility. At the time, all the thinking and training had been directed at their attacking major fleet units such as battleships, so this arrangement was in that context quite logical. However, they were also used independently by Keyes to form a close blockade against German warships in the Heligoland Bight, and even in the Baltic.

The work of the Northern Patrol, whose role as a blockading force against trade was to be vital in the eventual victory over Germany, is covered in Chapter 7. Despite its being in much the same waters, it actually impinged little on the war that was about to be fought in the North Sea, although the Grand Fleet did prevent the *Hochseeflotte* from interfering in its work.

There was one other force that was of great importance, which has been left to the last because it illustrates an important feature of the naval war conducted by Britain. The forces in the Straits of Dover originally came under the command of the admiral of patrols. The rapidly changing picture on land, with the German army advancing along the Belgian coast, forced a change. Churchill, in a quixotic move, absented himself from the Admiralty and attempted personally to rally the defence of the key port of Antwerp, to no avail. Of greater concern was the German occupation of the ports of Ostend and Zeebrugge. Based there, German naval forces, submarine and surface, could and did easily threaten British communications and the supply of its army in France. Showing what was to be characteristic flexibility, the Admiralty changed the command structure (as it was to do again during World War II, in very similar circumstances) and created a separate command under Rear Admiral Horace Hood. He commanded the Dover Patrol and was senior naval officer, Dover, initially with four cruisers and the 6th Destroyer Flotilla, comprised of 24 destroyers, 13 submarines and various smaller vessels. As Britain was responsible for the English Channel, four French destroyers came under his command when the French requested assistance to defend their Channel ports. It has been said that when Admiral Hood commanded operations from the *Intrépide* it was the first time that a British admiral had flown his flag in action in a French warship. One forgotten aspect of the naval war is that not a single British soldier was lost to enemy action crossing the Channel at any time during the war, despite the proximity of German naval bases to the Straits of Dover.[26]

Change at the top

There was a major change in the personnel at the top of the navy quite early on. Admiral Battenberg, the First Sea Lord, was of German birth. At the outbreak of war there was a great public antipathy to any and all things seen as being German; there were even incidents of dachshunds being attacked (by humans!). In the early days, and actually for some time thereafter, the navy was not seen to be doing anything, and Prince Louis of Battenberg became a target of public criticism because of his birth. He might well have survived this; however, despite his brilliance in peacetime, he proved not to be fit for the role in wartime. Churchill himself was not immune to criticism; his attempt to prevent the German army from occupying Antwerp with a handful of part-trained troops of the RND occasioned comment both public and private. By the end of October, in the interests of the navy, broken in spirit, Battenberg resigned, stating in his resignation letter: 'I have lately been driven to the painful conclusion that at this juncture my birth and parentage have the effect of impairing in some respects my usefulness in the Board of Admiralty.' His son was then a cadet at the Royal Naval College, Osborne, and understandably was deeply affected by this episode. It played a part in firing his ambition and his change of name to Mountbatten.[27] At this point Churchill, having decried Callaghan as being too old at 63 to be the commander in chief of the Grand Fleet, threatened resignation if Fisher, aged 73, was not recalled from retirement as First Sea Lord. So Fisher returned to the Admiralty.[28]

At the outbreak of war, the Royal Navy was at a high state of readiness and, in Churchill's words, 'looked for open battle on the sea. We expected it and we courted it', but Germany did not. There was not to be a fleet action for many months, but the initial months of the war were far from quiet at sea, even if, with the passage of years, they have been overshadowed by the war on land.[29]

CHAPTER 3

Towards Armageddon

E'en now their vanguard gathers,
E'en now we face the fray —

Rudyard Kipling,
'Hymn before Action'

The *Goeben* and the *Breslau*

The political situation in the Mediterranean in August 1914 was complex. Austria-Hungary, then a significant sea power, had been in large part responsible for the outbreak of war. However, while Germany was at war with Britain, initially Austria-Hungary was not. It had a small but modern fleet, mainly based at Pola, a port on the Eastern side of the Adriatic. Italy was part of the Triple Alliance with Germany and Austria-Hungary, but was not bound to come to war unless one of its partners was attacked. Its erstwhile allies saw it as being likely to remain neutral or even, as eventually happened, to join the Triple Entente comprised of France, Russia and Britain. Of the other littoral powers most were at this stage neutral, except of course Serbia, Austria-Hungary's *casus belli*.[1]

In the years leading up to the war, Germany had spent a lot of diplomatic effort on Turkey, the heart of the Ottoman Empire, not least in involving Turkey in the construction of its Berlin–Baghdad railway. As war loomed, British shipyards were building two Turkish battleships, which were very close to completion. In fact when war broke out there were Turkish sailors in England ready to sail them to Turkey. These ships were something of a national project; women had even sold their hair to help pay for them, although there had also been more orthodox contributions. Before war broke out, on Churchill's orders, the two ships were requisitioned for the Royal Navy as HMS *Agincourt* and HMS *Erin*; armed troops were used to prevent Turkish sailors from boarding them. The fact that two partly completed Chilean battleships were also taken over at much the same time, one ultimately to be completed as the aircraft carrier HMS *Eagle*, did nothing to reduce Turkish anger.[2]

The most important factor for the entente allies as war came was France's movement of troops by convoy from its North African possessions to metropolitan France. Critical was the transfer of the 19th Army Corps, about 80,000 men, from Africa to France. To transport them the French had an assortment of ships, including one Dreadnought, six pre-Dreadnought battleships and six armoured cruisers. The Austrians had two Dreadnoughts and three pre-Dreadnoughts based at Pola. The Germans also used Pola as a base for the operations of a battlecruiser, the SMS *Goeben*, and a cruiser, the SMS *Breslau*, under Admiral Wilhelm Souchon. The two German ships had been in the Mediterranean since 1912 and were in need of maintenance. As war seemed increasingly likely, this was Souchon's highest priority. At that point his ships were at sea, so he took them to Pola to undertake essential repairs to their boilers. His second priority over the next few days was fuel: coal for his two ships.[3]

Opposing them, the Royal Navy had three battlecruisers (*Inflexible, Indomitable* and *Indefatigable*), four large armoured cruisers, four light cruisers and a flotilla of 16 destroyers. The British commander in chief was Admiral Berkeley Milne (who rejoiced in the nickname 'Arky-Barky') and his deputy was Rear Admiral Troubridge, bearer of a famous naval name: Troubridge's great-grandfather had been one of Nelson's 'band of brothers'.

At this point the Admiralty became very involved with the developing situation. The Admiralty, with Churchill as its political head, was a unique organization in the British government. It was a department of state, a ministry – indeed in financial terms it was the biggest one. Crucially, and unlike the War Office, the department of state at the head of the army, the Admiralty was an operational headquarters; it ran the navy at sea, using wireless.[4]

When he became First Sea Lord, recognizing that the main threat was Germany, Admiral Fisher set about concentrating the Royal Navy in home waters. He had also driven centralization of command, writing in 1904:

telegraphy has been enormously developed, hence transmission of orders, and mutual conference of thought enormously bettered. Instead of a number of isolated squadrons acting under different heads, all actuated from different views of the same state of conditions, each independent of the strategy pursued by the other, we will have a co-ordinated whole. [...] united divisions under one Master Mind [sic] are infinitely stronger than a number of isolated squadrons [...] An enemy [...] will be the objective of the "Mind" in command.[5]

Churchill as First Lord took over the reins of the Admiralty in a way that some thought usurped the role of the First Sea Lord, and he was very much involved in making operational decisions. There is no doubt that he saw himself as the 'Mind' envisioned by Fisher.

Viewed in purely military terms, the highest Allied priority in the Mediterranean was to protect the troop convoys, and such was the first part of an instructing signal sent by Churchill. His initial order was followed by a stream of signals, some contradictory, giving Milne a series of rapidly changing tasks, which included watching the mouth of the Adriatic in the event of Austria and Britain going to war and shadowing the German ships; and all the time he was to husband his force and not to engage a superior enemy. Milne did not need to be told, for example, that 'the speed of your squadrons is sufficient to enable you to choose your moment'. It would have been much better if Churchill had given Milne the priorities for his command and let him get on with it. However, this was the first naval war in which it was possible for the Admiralty to interfere in real time. Perhaps, for a naval amateur, the temptation to micromanage was just too great.[6]

Goeben and Breslau on the move

Souchon left Pola and attempted to coal at Brindisi and then at Messina in Sicily, but the Italians were uncooperative. The Germans took all the coal they could find from German merchantmen in the port and went in search of the French troop convoys. They steamed to North Africa and almost immediately on the outbreak of war with France bombarded the ports of Philippeville (modern Skikda) and Bone in French North Africa. Having done so, they set out to return to Messina. Britain was not yet at war with Germany. Concerned that they might attempt to leave the Mediterranean,

Fig. 3.1. The SMS *Goeben* in the Turkish navy as the *Yavuz Sultan Selim*.

Churchill ordered two battlecruisers to Gibraltar to guard against this possibility. En route they encountered Souchon's force and turned to shadow.
Unfortunately they could not keep up with the German ships and, having
lost contact, had to detach to take on coal. Souchon was now able to coal
at Messina and was faced with a difficult decision. Austria was not yet at
war with France, and Turkey was havering between neutrality and declaring war on the side of Germany. Troubridge remained at the mouth of the
Adriatic with four armoured cruisers and eight destroyers. Souchon left
Messina for the Aegean, but feinted at entering the Adriatic. However,
he was shadowed by the light cruiser HMS *Gloucester*, which remained in
sight but out of gun range. She continually reported Souchon's position,
course and speed despite German efforts to jam her wireless signals. As a
result, Troubridge left his station and steamed south to intercept Souchon.

Goeben was armed with ten 11-inch guns and the *Breslau* had twelve 4.1-
inch guns. Souchon's ships were faster than Troubridge's armoured cruisers, although to conserve fuel they were steaming slower than Troubridge.
However, his guns were heavier and had a longer range than the twenty-two
9.2-inch guns of the HMS *Defence*, HMS *Warrior*, HMS *Black Prince* and HMS
Duke of Edinburgh. Troubridge did have eight torpedo-armed destroyers in
company. However, Troubridge was mindful of the direction he had been
given from Admiral Milne that he was not to engage a superior enemy.

He was aware that he would not come upon the German ships before full daylight, and thus would lose the advantage of darkness. He discussed matters with his flag captain, Captain Wray, and as a result broke off the chase. *Gloucester*, armed with a mixture of 6-inch and 4-inch guns, continued to shadow and actually engaged the *Breslau* in an attempt to slow her down. *Gloucester* hit her without effect and was engaged by the *Goeben* before being given a direct order by Milne to break off the pursuit.[7]

An Admiralty court of inquiry into Troubridge's decision was followed by a court martial. He was acquitted but never served at sea again. It is difficult to know what went through his mind when he decided to break off the chase. Certainly inconsistent orders and directions must have played a part.

Whether Troubridge could have sunk or damaged the *Goeben* is a moot point. Within months, two British battlecruisers were to fire two-thirds of their ammunition in sinking two German armoured cruisers during the battle of the Falklands; the *Goeben*, when fighting Russian ships, proved able only to fire at two targets simultaneously, which suggests that Troubridge, with four armoured cruisers and a destroyer flotilla, might have had a significant chance of victory. The effect of the events echoed down the years; in 1939 Commodore Harwood engaged a German pocket battleship, *Graf Spee*, the equal of a battlecruiser, with one heavy and two light cruisers at the battle of the River Plate. After that action the then First Sea Lord, Admiral Pound, alluding to the 'Troubridge affair', wrote to Commodore Harwood, saying: 'Even if all your ships had been sunk you would have done the right thing.' One suspects that the World War II destroyer HMS *Troubridge*, whose name would have been approved by Admiral Pound, was named for his great-grandfather.[8]

Once war had broken out, the Admiralty's main focus shifted to the North Sea, and it was here that the first significant surface actions were to take place.

The North Sea

Britain declared war on Germany at midnight on 4 August 1914. While the pursuit of the *Goeben* and the *Breslau* occupied the minds of the Admiralty, the Harwich Force went about its business. A routine had already been established: a 'duty' patrol flotilla would leave Harwich at about 04:00

THE GERMAN MINE LAYER "KOENIGEN LUISE," SUNK IN THE ACT OF LAYING MINES BY H.M. SHIPS
"LANCE," "LARK," "LINNET," "LANDRAIL," AND "AMPHION" OFF HARWICH. 5TH AUGUST, 1914.
ABRAHAMS & SONS, DEVONPORT. 674

Fig. 3.2. The German minelayer SS *Königin Luise*, sunk
in the first naval action of World War I.

Fig. 3.3. HMS *Lance*, which fired the first British shot of the war.

and cross the North Sea to relieve the previous day's patrol. It became a tradition of the Harwich Force that their commodore, Reginald Tyrwhitt, would wish every patrol 'good hunting' on sailing. By dawn the ships – light cruisers and destroyers – would be positioned in a line from Heligoland to the mouth of the River Ems to watch for an enemy patrol or forces moving along the coast with the intention of interdicting the movement of the British Expeditionary Force (BEF) across the Channel.[9]

On the morning of 5 August, the two destroyers HMS *Lance* and HMS *Landrail*, part of a flotilla attached to the cruiser HMS *Amphion*, came upon a ship that was painted like a Great Eastern Railway steamer. However, she was spotted to be dropping objects over the side which looked like sea mines.

The destroyers ordered her to stop, whereupon the ship opened fire on the destroyers. *Amphion* and the destroyers in turn rapidly sank the ship, which turned out to be a German minelayer, the SS *Königin Luise*. *Lance* may well have fired the first British shots of the war.[10] The survivors from the *Königin Luise* were rescued, but ironically most of them were to be killed the next day when the *Amphion* hit one of the mines their own ship had laid earlier.[11]

The battle of Heligoland Bight

Over the next few days there were numerous small skirmishes, culminating in the battle of Heligoland Bight on 28 August. Commodore Keyes's submarines based at Harwich had been operating in the Heligoland Bight, watching for German surface units. With Commodore Tyrwhitt he proposed to the Admiralty that his submarines should lie on the surface and draw surface units out onto Tyrwhitt's Harwich Force. They asked for the Grand Fleet to be available in support, including Admiral Goodenough's 1st Light Cruiser Squadron (1LCS). Admiral Sturdee, as the chief of staff at the Admiralty, approved the positioning of only two battlecruisers (*New Zealand* and *Invincible*) and the *Cressy* armoured cruisers (including the *Hogue*) between 40 and 100 miles to the west of the operation. Admiral Jellicoe got to hear of the operation and proposed that he have the Grand Fleet at sea in support. Sturdee signalled: 'Cooperation by battle fleet not required. Battle Cruisers can support if convenient.' Jellicoe took this as allowing

Fig. 3.4. HMS *Arethusa*, in which Commodore Tyrwhitt flew his
commodore's broad pennant at the battle of Heligoland Bight.

him some freedom of action and sailed three of Beatty's battlecruisers with
Commodore Goodenough commanding the 1st Light Cruiser Squadron.
He then followed with four battle squadrons of the Grand Fleet, only telling the Admiralty once he was at sea. Thus only once the operation was
actually underway did Tyrwhitt become aware that Goodenough's cruisers
were taking part in the operation, as were Beatty's battlecruisers. Keyes,
who was superintending his submarines from HMS *Lurcher*, however, was
unaware of this; indeed he and his submarine commanding officers had
been told before sailing that the cruisers *Arethusa* and *Fearless* were the
largest British ships that would be in the Bight.[12]

Tyrwhitt had been hampered in controlling his 30-knot destroyers from
his flagship the *Amethyst*. Despite being the first British turbine-engined
cruiser, she could only make 22 knots, so at the earliest opportunity
Tyrwhitt shifted his broad pennant to the brand-new *Arethusa* and commanded his force from her. *Arethusa* could make 29 knots, but she had
only commissioned a fortnight earlier and was far from a fully worked-up
fighting unit.

The Germans had been expecting the British to come into the Bight, but were surprisingly unprepared. On a day when, as is common in the North Sea, there were a lot of mist and fog patches, a confused engagement followed. The charts which record the tracks of the individual ships during the battle 'are possibly the most difficult to follow of any battle during the war'.[13]

Fortunately for the British, because it was low water in the Jade River, where the German fleet was at anchor, the German big ships and the battlecruisers could not cross the sandbar at the mouth of the river and get to sea. As a result, what followed was initially a cruiser and destroyer action. Early on the *Arethusa* was engaged in a single-ship action with the SMS *Frauenlob*. Two of *Arethusa*'s guns now jammed. Gun-jamming was usually the result of a drill error and inadequately practised gun's crews, evidence of *Arethusa* having only commissioned very recently. She was hit 15 times but gave better than she got, and the *Frauenlob* broke off the action. Then the German destroyer *V-187* encountered eight British destroyers and was rapidly sunk.

There were some moments that bordered on farce. Keyes, on encountering light cruisers, assumed that they must be German. Accordingly, he signalled Tyrwhitt for assistance. Tyrwhitt passed on the request to Goodenough. It transpired that Goodenough was being asked to assist by engaging himself! The lack of information threatened tragedy when the British submarine *E6* fired two torpedoes at the cruiser wearing Goodenough's broad pendant, HMS *Southampton*. Luckily they missed. Understandably, on sighting a periscope the *Southampton* attempted to ram her attacker, a presumed German submarine, and with the submarine at periscope depth she could well have hit her, but fortunately *E6* was able to dive below the cruiser.

The Germans now sent more and more destroyers and cruisers into the battle. They arrived singly rather than in a concentrated mass, and a further series of confused actions took place, complicated again by ships being misidentified. The German light cruiser SMS *Mainz* was sunk, but increasingly the British ships were getting the worst of the engagement when Beatty intervened with five battlecruisers, *Lion*, *Queen Mary*, *Princess Royal*, *Invincible* and *New Zealand*. Until the tide would allow the German battlecruisers to get to sea, the Germans were totally outgunned. The

British sank three light cruisers and a destroyer, and damaged three more cruisers. The Royal Navy inflicted 1242 casualties while sustaining only 75 with the loss of no ships, although the *Arethusa* had to be towed home by the *Hogue*.[14]

While the battle of Heligoland Bight was a victory, it could easily have been a disaster. For the first time, but not the last, Sturdee's judgement was called into question. The next major error he made was the disposition of the 'live-bait squadron', and the resulting loss of the *Hogue*, *Aboukir* and *Cressy* in September following the battle of Heligoland Bight went some way to his incurring Fisher's dislike, which contributed to his leaving the Admiralty. He was soon to leave in order to take command of the squadron charged with avenging the defeat at the battle of Coronel at the hands of Admiral von Spee and the German Far East Squadron.

The German Far East Squadron

As part of its attempt to build an empire like the other European powers, in 1897 Germany had a Far East Squadron that was actually based in the British colony of Hong Kong. Admiral Tirpitz personally established the only overseas naval base the Imperial German navy ever had. After visiting the Far East, Tirpitz settled on Tsingtao (modern-day Qingdao) on Kiaochow (Jiaozhou) Bay on the Yellow Sea, which was first captured and occupied by German marines and then in 1899 leased from the Chinese government for 99 years. It became a little corner of Germany, and the Tsingtao Brewery established there still makes beer for export to this day. While it was a colony that produced coal for export from its hinterland, it had been established and was maintained as a base for the East Asia Squadron.

In 1914 the squadron was commanded by Vice Admiral Graf Maximilian von Spee. He was a gunnery specialist and under his command he had two armoured cruisers, *Scharnhorst* and *Gneisenau*. They had twice won the Kaiser's Cup as the best gunnery ships in the navy. Like the three light cruisers he had under command, the SMS *Emden*, SMS *Leipzig* and SMS *Nürnberg*, they were manned by long-service officers and ratings; these were ships at the top of their game, but hoping to be home for Christmas. When war threatened, his two major units had already embarked on a three-month cruise to a variety of ports. Von Spee overrode instructions from Berlin given to *Nürnberg* and concentrated his squadron at Ponape, 400 miles east of Truk. He was accompanied by an armed merchant cruiser (an armed passenger liner), the SS *Prinz Eitel Friedrich*, and eight supply ships. With Japan joining the Allies, he realized that to return to Tsingtao was impossible. So when the British cruiser HMS *Minotaur* destroyed the German radio station in the Caroline Islands, he temporarily detached the *Nürnberg* to Honolulu to advise the naval staff in Berlin by cable of his intention to head for South America, specifically Chile. This was because there was significant German influence and infrastructure there, in particular access to the all-important coal for fuel. At the same time he detached the *Emden* and *Prinz Eitel Friedrich* to undertake commerce raiding.

At this point, the British priority was to ensure the safe passage of troop convoys from Australia and New Zealand proceeding to Europe. To protect them in the Pacific and East Indies the British and Dominion (i.e. Australian) navies had, split among three squadrons, a modern battlecruiser, two pre-Dreadnought battleships and 17 cruisers of various types. They were far from efficient units: HMS *Triumph* had been demobilized before war broke out and she eventually went to sea with her small naval crew supplemented by over 100 volunteer soldiers. However, the focus shifted once Japan entered the war and it became obvious that Spee was heading for South America. British trade in the South Atlantic was of major importance. Normally Germany kept one cruiser in the Western Atlantic, but the war came as the *Dresden* was handing over to the *Karlsruhe*, so there were two. If they joined up with Spee's squadron, there would potentially be a very serious threat to Allied trade routes.

Initially the Admiralty was unaware that the *Emden*, commanded by *Fregattenkapitän* Müller, had been detached by Spee to operate independently. Thus it was a surprise when she appeared in the Bay of Bengal, where normal peacetime shipping had resumed after the initial excitements following the outbreak of war. In five days the *Emden* sank six ships and captured two to use as supporting colliers. Just over a week later, on 22 September, the *Emden* closed the Indian coast at Madras and bombarded its oil storage tanks. She then went on to sink or capture another six ships, landing their captured crews in one of them. She then raided Penang, sinking a Russian light cruiser and a French destroyer. In total *Emden* sank 16 steamers, totalling 70,825 tons. She then went on to the Cocos Islands, intending to destroy its wireless and cable communications. A wireless warning led to the Australian cruiser HMAS *Sydney* being detached from the escort of a nearby convoy; on closing the Cocos she encountered the *Emden* and sank her on 9 November.

Meanwhile, the converted liner *Prinz Eitel Friedrich*, operating mainly in the Pacific and South Atlantic, sank 11 ships, mostly sailing vessels, before being voluntarily interned in the United States (then neutral) in March 1915.

There was a further German cruiser in the Indian Ocean, the SMS *Königsberg*, commanded by *Fregattenkapitän* Max Looff. Continually short of coal due to adept Admiralty control of local shipping, she only sank a single merchant ship and drove the HMS *Pegasus*, a cruiser, ashore. *Pegasus* had been repairing machinery defects in Zanzibar. *Königsberg* then took refuge in the delta of the Rufigi River in German East Africa, where she was found by the *Dartmouth* and blockaded to prevent her from going back out to sea. Finding her was only half the battle. She was too far upriver for accurate gunfire to reach her from the sea. Eventually, after failed attempts to bomb her from the air, Fisher sent two ex-Brazilian monitors, HMS *Severn* and HMS *Mersey*, out from Malta. These each had a twin 6-inch gun turret. Painted green for camouflage, they entered the estuary despite German shore guns and commenced bombarding the *Königsberg* on 6 July 1915. They used an aircraft to spot their fall of shot. While they did achieve some hits, tidal considerations forced them to withdraw until 11 July, when they returned and sank her.[15]

While the raiding cruisers achieved little in terms of material damage, they did cause significant disruption to trade, and it took a lot of effort to locate and sink them. Thus the cruisers, both those deployed specifically to raid and those detached by Spee to do the same, caused disproportionate disruption and effort to the Allies.

In the meantime, the colony at Tsingtao was instructed personally by the Kaiser to fight to the end. A largely Japanese naval and army landing force, supported by two battalions of British troops, *Triumph* and a destroyer soon overcame the German resistance, which surrendered on 7 November.

By the time Tsingtao fell, the Far East Squadron had inflicted a defeat on the Royal Navy at the battle of Coronel. At the outbreak of war, Rear Admiral Sir Christopher Cradock was the commander in chief of the North

Fig. 3.5. HMS *Good Hope*, Rear Admiral Cradock's flagship at the battle
of Coronel. Built in 1901, her main armament was two 9.2-inch guns.

American and West Indies station. It soon became apparent that the major
threat was further south, with the German cruiser *Dresden* operating as a
raider off the mouth of the Amazon. The Admiralty created a new South
American station, under Admiral Cradock, primarily to find and sink the
Dresden.

Under his command, Cradock had two obsolete armoured cruisers,
the *Monmouth* and *Good Hope*, flying his flag in the latter. Both had been
Third Fleet ships. When the war began they were 90 per cent manned by
reservists, boys and cadets. Worse still, they had been allowed only four
rounds per gun as practice ammunition on commissioning. Cradock had
in addition two armed merchant cruisers, the HMS *Otranto* and the HMS
Carmania. The latter was detached to reconnoitre Trinidad Island (not the
West Indian Island of the same name; this one lies about 500 miles east
of Rio de Janeiro). There, on 14 September, she encountered a German
armed merchant cruiser, the SMS *Cap Trafalgar*. Despite being severely
damaged herself, *Carmania* sank the *Cap Trafalgar*. Meanwhile, Cradock's
squadron steamed south to be joined by HMS *Glasgow*.[16]

Fig. 3.6. HMS *Monmouth*. Built in 1901, she was armed with
fourteen 6-inch guns and like the *Good Hope* was hopelessly
outgunned by both *Scharnhorst* and *Gneisenau*.

Fig. 3.7. HMS *Canopus* was completed in 1897. She had four
12-inch guns and, when new, could make nearly 19 knots. By
1914 her engines were subject to repeated breakdowns.

Churchill, the Admiralty and Cradock now started to worry about Spee. The Admiralty sent Cradock reinforcements, HMS *Defence* and HMS *Canopus*, and a wordy signal. It bore the hallmarks of the signals sent to Milne and Troubridge, being complex and contradictory. Cradock was firstly ordered to leave sufficient force in the Atlantic to cope with the two German cruisers *Dresden* and *Karlsruhe*, which were then thought to be in that ocean. He was also to concentrate his force to meet Spee, who was believed to be crossing the Pacific Ocean.

Of the two ships sent to reinforce him, *Canopus* was another Third Fleet ship. She was a pre-Dreadnought battleship with four 12-inch guns. However, they were outranged by Spee's guns. Her elderly engines were unreliable and at best she was 6 knots slower than the German ships. Her two gun turrets 'were in the charge of Royal Naval Reserve Lieutenants who, before the war, had never stepped inside a battleship gun turret'. This was the ship that Churchill described to Cradock as 'a citadel around which all our cruisers in those waters could find absolute security'. On 12 October Churchill wrote: 'They [*Scharnhorst* and *Gneisenau*] are our quarry for the moment, we must not miss them.' How Cradock was to intercept and destroy a faster squadron if he included *Canopus*, Churchill did not make clear. Then the Admiralty changed its mind about sending *Defence* to join Cradock, because of a change in the perceived threat of Spee coming into the Western Pacific. However, the Admiralty did not at this stage tell Cradock she was not coming to him. His orders, never rescinded, were to find and destroy the Germans. With the agreement of the Admiralty, he sailed from the Falklands in the *Good Hope* to join up with *Monmouth*, *Glasgow* and *Otranto*, which he had already dispatched to the west coast of South America along with *Canopus*, which was to follow as best she could when she had completed repairs to her machinery.[17]

The Admiralty, in response to a suggestion from Cradock, who was concerned that Spee might be able to bypass him and attack the trade routes in the South Atlantic, formed an East Coast Squadron and allocated *Defence* to it. Rather than put both under one admiral, the East Coast Squadron was under Rear Admiral Stoddart. Cradock, being senior to Stoddart, ordered Stoddart to release *Defence* to him, but his orders were countermanded by the Admiralty.

Thus Cradock fought Spee with two relics (*Monmouth* and *Good Hope*) saved from the breaker's yard by the outbreak of war – an armed merchant cruiser, the *Otranto*, and one modern light cruiser, the *Glasgow*. Lumbering up behind him was another barely seaworthy relic, the *Canopus*, whose forward gun turret flooded in a head sea.

The *Glasgow* went into the port of Coronel on 30 October to pass signals to the Admiralty using the cable station. While there she heard increasingly powerful German *Telefunken*, or radio traffic, suggesting that Spee was getting closer. She joined the other ships at sea. Cradock then set up a patrol line moving north, looking for the German ships. Because he was apparently expecting to encounter a light cruiser, the *Otranto* was kept in company, which was to be a fatal mistake. When the British and German squadrons came into contact the British reversed course, and both squadrons steamed south almost parallel to each other, about 1,500 yards apart. Cradock was to seaward, and because all his ships except *Otranto* were faster than the German ships, he did not have to accept action; he could have turned away, but he felt he could not run and leave *Otranto*, and of course his orders were to engage the two German ships. He did have the initial advantage that the Germans would be firing into the setting sun, making life difficult for their gun-layers and range-takers. Once the sun had set, however, his ships would be silhouetted against the twilit sky. The sun soon set and the action lasted less than an hour. Both *Monmouth* and *Good Hope* were sunk with no survivors. *Glasgow* was hit five times. *Otranto* made good her escape into the gloom. *Scharnhorst* was hit twice by shells that failed to explode and *Gneisenau* was hit four times, sustaining three minor casualties, the totality of the damage and casualties sustained by the Germans.

Glasgow fell back to the south to join up with *Canopus*, and both then set off for the Falklands. The further away they were from the German ships, the less the effect of German radio jamming, and so *Glasgow* was able to report to the Admiralty the result of the encounter. At this point it appears that the Admiralty at last realized that modern ships were needed to fight modern ships; it was insufficient to count gun sizes and ignore the machinery aspect or the state of the personnel and their training. Nonetheless, Churchill continued to believe that Cradock should have remained in company with *Canopus*, where he would have been safe.

Fig. 3.8. The memorial to the battle of the Falklands
in Port Stanley, Falkland Islands.

The Admiralty then strengthened Admiral Stoddart's squadron; now he had a total of four armoured cruisers and two light cruisers under his command. *Canopus* was ordered to ground herself as a coastal defence ship in Port Stanley harbour in the Falklands. Most importantly, three battle-cruisers were detached from the Grand Fleet, first *Invincible* and *Inflexible* to go to the South Atlantic, and five days later *Princess Royal* to the West Indies in case Spee went north and transited the Panama Canal. The first two stored and repaired essential defects at Plymouth and sailed for the South Atlantic on 11 November, *Invincible* still carrying dockyard workers. They were under the command of Vice Admiral Sturdee, until then the chief of staff at the Admiralty. For Fisher, this choice solved another problem: he could not stand Sturdee and had a low opinion of him professionally.[18]

Gradually picking up ships on his way south, Sturdee arrived in the Falklands on 7 December. His squadron now comprised the two battle-cruisers, and the cruisers *Carnarvon*, *Glasgow*, *Kent*, *Bristol*, *Cornwall* and the armed merchant cruiser *Macedonia* – and of course *Canopus* had been waiting for him in the Falkland Islands. His ships were coaling and/or repairing their machinery when Spee arrived, intending to bombard the colony on 8 December. One of Spee's officers reported sighting tripod masts, which could only indicate the presence of British Dreadnoughts, but this was discounted. Only when the first shells from the *Canopus* fell did the German ships turn away to the south-east. Sturdee directed *Bristol* – delayed by reassembling an engine – and *Macedonia* to capture Spee's colliers while he signalled 'general chase' to the other ships. This order allowed ships' independent action in pursuit of a fleeing enemy. *Inflexible* opened fire at the then extreme range of 16,500 yards. Throughout the action, Sturdee endeavoured to fight at long range, outside the range of the German ships but within his own. In an attempt to allow his light cruisers to escape, Spee split his force. Sturdee did likewise, sending his light cruisers to follow the Germans. Thus the battle became one between the heavy ships and a series of small-ship actions. The battlecruisers' shooting was hampered by a combination of long range, wind, and funnel and gun smoke. The result was that the British ships did not shoot well, but the weight of fire told. *Scharnhorst* was the first to sink, followed by *Gneisenau* after a further 90 minutes. There were 176 survivors. Of the cruisers only *Dresden* managed to outrun her pursuers. *Glasgow* and *Cornwall* sank the *Leipzig*, and *Kent* the

Nürnberg. There were few German survivors and very few British casualties. Admiral Cradock and the companies of *Good Hope* and *Monmouth* had been avenged.

Immediately after the battle the British gunnery performance attracted unfavourable professional comment. However, as the war progressed it came to be recognized that wartime shooting was very different from peacetime practice. In particular ranges smoke, a ship's speed and the consequent vibration affecting optical instruments meant that British ships achieved about 5 per cent of hits in expending two-thirds of their ammunition.[19]

The *Dresden* outran her pursuers and returned to the Pacific side of South America. Despite rumours from ashore passed on by various British consuls, she was not found for months, during which time she was hiding in the inlets and passages around the tip of South America, repeatedly breaching Chile's neutrality. Following a distant sighting by the *Kent*, which had insufficient fuel to pursue, Captain Luce in *Glasgow*, in company with the armed merchant cruiser *Orama*, located her in the Juan Fernández group of islands, 600 miles off Chile. Joined by the *Kent* after she had coaled, they closed on the islands.

They encountered the *Dresden* in Chilean territorial waters. The German ship was in breach of Chilean neutrality, but Chile had been in no position to enforce it. Notwithstanding diplomatic niceties, Captain Luce opened fire, and within six minutes *Kapitän* Lüdecke struck his colours indicating that he surrendered, and then scuttled his ship. Afterwards Chile protested to both Germany and Britain about the breaches of her neutrality. Britain's attitude was that she was legally in the right, invoking the doctrine of 'hot chase' and arguing that regrettably Captain Luce had had no alternative. The Chilean government accepted this as an apology.[20]

With the neutralization of most of the German surface raiders, Admiralty attention was now again brought back to the North Sea. One of the arguments deployed before the war by the protagonists of close blockade of the German coast was that a distant blockade would allow German ships to launch attacks on the British coastline. At the time, no one envisaged that the Germans would attack civilian targets. The German adoption of a policy of *Schrecklichkeit* (literally, 'frightfulness' or 'terror') to restrain the civilian populations of the countries they marched through showed that they were prepared to attack civilian targets from the outset of the war.[21]

The *Hochseeflotte* throughout the war had a smaller number of Dreadnought warships than the Grand Fleet. The Kaiser instructed that his ships were not to fight superior numbers, so successive German naval commanders repeatedly tried various stratagems to lure a portion of the Grand Fleet into a position where it could be destroyed. As Beatty's battlecruisers were much faster than the rest of the Grand Fleet and stationed much further south, they were the likely targets. Hence on 3 November the German navy started what was to be an intermittent campaign of bombardments of British coastal towns.

A force commanded by Admiral Hipper, comprising the battlecruisers SMS *Seydlitz*, SMS *Moltke*, SMS *Von Der Tann* and SMS *Blücher*, was dispatched to bombard the port of Yarmouth. They were accompanied by four light cruisers that would lay mines and were to be supported by a sizeable portion of the *Hochseeflotte*. The hope was that this attack would draw units of the Grand Fleet into a trap. They sailed on 2 November and in the early hours of 3 November the German battlecruisers shelled Yarmouth beach (which was of as much military value as any other land target in the area). As they withdrew they failed to sink a minesweeper and a destroyer that they ran into and lost the cruiser SMS *Yorck*, which hit a mine. The British lost the submarine *D5*, which also hit a mine.

Following the Battle of the Falklands, the Germans were now aware that some battlecruisers had been detached to the South Atlantic, thus weakening the Grand Fleet. This presented them with an opportunity. At about the same time, following the Russians' capture of a German code-book from the *Magdeburg*, which was then passed to Britain, Room 40 broke into the German naval codes. Within six weeks they were able to provide useful warning of the next planned bombardment of towns on the east coast. On 14 December the Admiralty sailed the battlecruisers under Beatty, a battle squadron under Vice Admiral Warrender, Tyrwhitt's light cruisers from Harwich and a screen of Keyes's submarines. What the Admiralty did not know was that the entire *Hochseeflotte* under Admiral von Ingenohl was in fact at sea. The first encounter was between screening destroyers and cruisers at 05:15 on 16 December. This was sufficient to worry Ingenohl, who had already exceeded his instructions from the Kaiser by taking his ships outside the Heligoland Bight and 'turned tail and made for home, leaving his raiding force in the air'. Had Ingenohl

GERMAN RAID ON THE ENGLISH COAST
AN ENGLISHMAN'S HOME IN RUINS AT SCARBOROUGH

Fig. 3.9. One of the results of the German bombardment
of Scarborough. A British propaganda picture.

continued for less than an hour, he would have come on a numerically
very inferior portion of the Grand Fleet and could hardly have helped but
win a significant victory.[22]

The German battlecruisers bombarded Scarborough and Hartlepool to
little military effect and made for home. There were fleeting encounters
between various units, but Hipper and his ships escaped. Commodore
Goodenough, commanding the *Southampton* and the 1st Light Cruiser
Squadron, came into action with the SMS *Stralsund*, one of Hipper's
screen, and then the *Strassburg* and *Graudenz*. He reported to Beatty that
he was in action with 'enemy cruisers'. Beatty initially let *Southampton* and
Birmingham leave his screen to support Goodenough, but when *Nottingham*
and *Falmouth* attempted to break away as well, not apparently realizing why
they had detached, Beatty ordered his staff that 'that cruiser' should rejoin
the screen of ships around his battlecruisers. His flag lieutenant, Lieutenant
Commander Seymour, not knowing which of the two cruisers with identical silhouettes then in sight Beatty meant, sent a signal by flashing light

(which was directional) with the order 'Light cruiser resume station'. It was sent without a call sign, which would have designated a given ship, and thus was read as recalling all the cruisers. Goodenough did as he was told by the senior officer and broke off the action. After an exchange of signals with Goodenough, Beatty realized what had happened. He then publicly upbraided Goodenough: 'When you sight the enemy, engage him [...] I cannot understand why, under any circumstances, you did not pursue the enemy.' Beatty was sufficiently angry to press Jellicoe, Goodenough's superior, to sack him, which he refused to do. Goodenough was to be present at the next encounter with a portion of the *Hochseeflotte* in the New Year.[23]

The battle of Dogger Bank

On 23 January 1915, Ingenohl allowed Admiral Hipper to reconnoitre Dogger Bank. He was ordered to sail that evening and to return the following evening. Nothing more was planned; the German fleet was not at full strength, as one of its battle squadrons was in the Baltic for training. Hipper took with him four battlecruisers, the large armoured cruiser SMS *Blücher*, four light cruisers and two destroyer flotillas. Taking *Blücher* was an odd decision; she was slower than the battlecruisers and far from as well armed.[24]

The Germans used wireless to pass on the instructions for the operation, and their signals were were intercepted and decrypted by Room 40. Forewarned, the Admiralty set a trap, comprising Beatty's battlecruisers led by his flagship *Lion* and Goodenough's light cruisers from the Grand Fleet in the north. In addition, positioned to the north-west were Tyrwhitt's three light cruisers and 35 destroyers from Harwich, with Vice Admiral Bradford commanding the 3rd Battle Squadron of eight *King Edward VII*-class pre-Dreadnought battleships. There were also three armoured cruisers from the 3rd Cruiser Squadron under Rear Admiral Pakenham.

There were some important differences between German and British gunnery material, doctrine and practice, which were to affect the battle that followed. The Germans did not intend to fight at long range; their preference was for battle ranges between 5,000 and 8,000 yards. The British were quite ready to fight at longer ranges as they had done at the battle of the Falklands. This reflected the way their guns were controlled. The

Germans used a central rangefinder as high up in the ship as practicable, and the range was passed to the individual gun turrets, who were then responsible for pointing the gun (training it) at the correct range (laying it). The Royal Navy had been moving from a similar system to a system of director control. A gunnery director high up in the ship was equipped with a rangefinder and the best gun-layers and -trainers in the ship. Information went from the director to the 'transmitting station'. This was deep within the ship behind the armour belt, reflecting its importance, and contained the 'fire control table'. This was a mechanical analogue computer, which made the necessary calculations to control gunfire over long ranges and passed the information to the gun turrets. When all the guns were loaded and correctly laid, they could fire one gun at a time from a given turret, termed a salvo, or all the guns together, known as a broadside, at the best moment. To ensure the guns remained pointing at the target despite the ship rolling and altering course, there had been very heavy investment in hydraulic power for gun turrets, which ensured very accurate and speedy control of their movements. At the time of the battle of Dogger Bank not all of Beatty's ships were fully equipped for director control (*Lion* was not).[25]

Aurora from the Harwich force was the first to encounter Hipper's cruiser screen, almost immediately followed by the sighting of large ships. At the same time the Germans intercepted wireless call signs, which to them suggested the close proximity of a battle squadron. Sensing the presence of superior enemy forces and mindful of his instructions from the Kaiser, Hipper turned for home. Goodenough now positioned himself to observe and report to Beatty. This meant he placed his ships out of the line of fire of either group, but reported first the composition and then the course and speed of the German force. This was the role of cruisers laid down by Jellicoe in his Grand Fleet Battle Orders. Goodenough also later passed on fire corrections when he observed *Tiger*'s shells falling beyond their intended target, an exemplary performance by a cruiser commodore.

On Beatty's staff was an American who had through personal connections managed to be commissioned as an RNVR lieutenant attached to Beatty's staff as a supernumerary. He made an interesting observation:

> Although during an action the Admiral and his staff are supposed to retire to
> the conning tower [...] the signal staff were still on the signal bridge and the

> rest of us including the Admiral [Beatty] [...] the flag lieutenants were all on
> the compass platform.'

Beatty did go into the relative security of the conning tower for a short period, but

> did not long remain in the conning tower. He was thoroughly enjoying him-
> self and did not like to waste his day in the cramped and crowded security of
> the conning tower and he and the flag lieutenant, the Flag Commander and
> Secretary were soon up on the compass platform.

Notably absent was the signal boatswain, a warrant officer specializing in signals. Thus Beatty was separated from his professional signallers; his flag lieutenant, Lieutenant Commander Ralph Seymour, was not signals-trained (the flag commander was an administrative post).[26]

A chase ensued, some of the British battlecruisers actually exceeding their design speed as they closed on Hipper's ships. Beatty's ships opened fire at 20,000 yards, a range not even contemplated before the war. It was important to ensure that each enemy ship was engaged. Standing orders issued to all ships set out which enemy ship to engage. Obviously there was a series of options, depending on how many enemy and friendly ships were involved. Thus on the day the standing orders would be supplemented by a 'fire distribution' signal. Beatty ordered that ships were to fire on 'corresponding ships' in the enemy line. An ambiguity arose; *Tiger* fired at the wrong ship and the *Moltke* was left to return fire undisturbed. To compound *Tiger*'s error, her shooting was poor. While she was a recently commissioned ship, she was fitted with director control that should have improved her shooting.

Very early in the chase Hipper's flagship, the *Seydlitz*, suffered a near-catastrophic hit from *Lion*. A hit on her aftermost turret set fire to charges in the hoist, coming up from the magazine to the guns. Magazine- and handling-room crews endeavouring to escape allowed the fire to spread to the next turret. One hit not only put two turrets out of action, but nearly sank the ship. Then *Blücher* was hit; she slowed and pulled out of the line, although continued fighting until her end.

The Germans now hit *Lion*. The first hit caused severe flooding, loss of a turret and also of all electrical power. This meant she could not communicate using wireless or by using flashing searchlights. *Lion* had also lost most of her signal halliards, the ropes used to haul signal flags up to give orders to the other ships; she now only had two. What followed can partly be blamed on that damage, but also on Beatty himself and his flag lieutenant, whose job it was to advise Beatty, as Lieutenant Pasco had advised Nelson on signal content at the battle of Trafalgar.

Because of the damage she had sustained, *Lion* slowed down. The other battlecruisers overtook her, despite the fact that Beatty did not hoist a flag known as 'Blue Burgee', which, according to the signal book, would indicate that 'his motions are not to be followed by the fleet'. At this point Beatty thought he saw a periscope on *Lion*'s starboard bow. Without telling anyone on the bridge the reason, he ordered an alteration of course to port by eight points (nearly 90°) away from the supposed submarine. After the battle there was some debate as to what Beatty should have done. Jellicoe suggested that he should have turned to starboard, towards the submarine, where he would have had a good chance of ramming it. Fisher questioned whether there had been a submarine in the area. It was certainly a case of 'periscopitis'. The problem was that Beatty did not have the signal made with a preceding submarine warning flag, so his captains did not know why he had ordered the change in course. Captain Pelly of the *Tiger* wondered if it was because of a minefield that Beatty knew about but others did not. As the other ships overtook him and rapidly drew away, Beatty ordered 'course north-east', which would have resumed the chase, and simultaneously to 'attack the rear of the enemy'. Apparently Beatty had wanted to make the famous signal 'engage the enemy more closely', but it was no longer in the signal book and the signal made to attack the rear of the enemy was 'the best Seymour could offer'. This is patent nonsense. The relevant page from the signal book is reproduced at Fig. 3.11. It can be seen that there were at least two obvious alternatives – 'AD' and 'AE' – that would have better conveyed his intent. He could also have used 'general chase', as Sturdee had done at the battle of the Falklands. However, he did not have the advice of a professional or competent signals officer immediately to hand.[27]

Flag signals were hoisted to be read and understood, but were only to be obeyed when hauled down – in the jargon, to be 'executed on the downhaul'. Unfortunately the two signals were hauled down at the same

Signal.	Attacking.
A A	Attack the Enemy by endeavouring to run them down.
A B	Attack the Enemy on their Starboard side.
A C	Attack the Enemy on their Port side.
A D	Attack the Van of the Enemy.
A E	Attack the Centre of the Enemy.
A F	Attack the Rear of the Enemy.
A G	Attack the Right of the Enemy. The Right of the Enemy is the part on the right hand when facing the Enemy's Centre.
A H	Attack the Left of the Enemy. The Left of the Enemy is the part on the left hand when facing the Enemy's Centre.
A I	Attack the Ships of the Enemy as indicated.
A J	Attack the Enemy with Whitehead Torpedoes.
A K	Attack the Torpedo Vessels of the Enemy.

Fig. 3.10. Page 41 of the signal book in use in 1915. Beatty signalled 'AF'. 'AD' or 'AE' would have conveyed his wishes far better.

Fig. 3.11. SMS *Blücher* sinking. She continued fighting to the end, despite being the focus of attention of the entire Battlecruiser Squadron, less HMS *Lion*.

time. Thus they were read as being a single order, not as the two separate orders that Beatty had intended. Admiral Moore, Beatty's deputy, realized that *Lion* was incapacitated and obeyed what he saw as Beatty's last order. The only enemy to the north-east was the unfortunate *Blücher*, and she was now to be the recipient of the undivided attention of four battlecruisers while the remainder of Hipper's ships escaped.

Beatty kept his personal flag flying, indicating that he was still in command, but he was unable actually to exercise this. He attempted to shift his flag, transferring to a destroyer, the *Attack*, and then to the *Princess Royal*, but it was too late to resume the chase: the gap was now too wide.

It is odd that Beatty's captains chose to obey what they saw as his orders rather than his widely known earlier injunction to Goodenough, who had, after all, obeyed what he saw as an order and been publicly condemned for doing so. As Fisher said of the episode, 'Any damn fool can obey orders.' It was all the more curious since Beatty had commanded the squadron for nearly two years; his captains by then should have been 'in his mind', as Nelson would have had it. Perhaps Beatty should have formally handed over command to Admiral Moore, which could have been done with a single flag hoist. Perhaps, as Admiral James Goldrick has commented, it would have been better if all of *Lion*'s signal halliards had been shot away.[28] Certainly the comment 'Nelson has come again', made to Churchill by a Battlecruiser Squadron officer after the battle, strains credulity.[29]

The next time the battlecruisers and Beatty, again with Seymour as his flag lieutenant, were to be in action was at the battle of Jutland, the naval 'Armageddon' that the service had been so eagerly anticipating.

Fig. 4.1. The memorial to the Royal Naval Division
on Horse Guards Parade in London.

CHAPTER 4

The Amphibious Navy

'E's a kind of giddy harumfrodite — soldier an' sailor too!

Rudyard Kipling
'Soldier an' Sailor Too'

At the outbreak of World War I the navy had for over 200 years had its own army, the Royal Marines.[1] They included infantry, the Royal Marine Light Infantry (RMLI), and the Royal Marine Artillery (RMA). While Fisher had regarded the army as a projectile to be fired by the navy, it wasn't averse to fighting on land. During the Victorian era, the navy landed sailors and marines to fight alongside the army on many occasions, and both Admirals Jellicoe and Beatty had participated.[2] During World War I, 40 per cent of its casualties were incurred in land fighting, and the only major naval war memorial in London dating back to the Great War is dedicated to the 'Royal Naval Division'.

It is a popular belief the Royal Naval Division came in to being almost by accident, at the whim of Winston Churchill, who dashed off to Antwerp at its head while the Germans advanced at the beginning of the war. Like many myths, there are elements of truth. The war book made provision for the Admiralty at the outbreak of war to form a Royal Marine Brigade made up of one RMA and three RMLI battalions. This 'advanced base force' was to seize and/or protect temporary naval bases necessary in support of the fleet or 'the provisioning of an army in the field'.[3]

At the outbreak of war, the Admiralty actually had more men than it immediately needed. These included men of the Royal Fleet Reserve (RFR), that is ex-Royal Navy personnel recalled at the outbreak of war, the Royal Naval Reserve (RNR), who were merchant seamen, and the Royal Naval Volunteer Reserve (RNVR), who can best be described as enthusiastic naval amateurs. To manage and employ these extra men usefully, the Admiralty created a division for land service. To oversee it an Admiralty standing committee was formed, chaired by the First Lord, Winston Churchill. On 16 August this formed the manpower into battalions, named for admirals. The battalions were formed into three brigades, two naval and one Royal Marine, each of four battalions. The officers were from the Royal Marines, the RN and the RNVR, with a sprinkling of ex-Guards officers supported by RN senior ratings. The 1st Brigade was comprised of Collingwood, Hawke, Benbow and Drake Battalions, the 2nd of Howe, Hood, Anson and Nelson Battalions, while the Royal Marine battalions were named for their base ports, Portsmouth, Plymouth, Chatham and Deal. When the division actually formed up, the RMA battalion was replaced by a fourth RMLI battalion, which meant that the division had no artillery of its own.

The two naval brigades formed up at Walmer and Betteshanger in Kent. Unfortunately the navy now started to withdraw the more experienced senior ratings for service in the fleet. Then the War Office requested that the Admiralty take over some of their surplus recruits from the north of England, to the extent that a Royal Naval Division manning depot had to be established at Crystal Palace. This eventually became the depot for RNVR personnel afloat or with the RND, functioning in the same way as the port divisions at Chatham, Portsmouth and Devonport did for the rest of the navy. The RND initially had no commanding officer, and while many of its RNVR officers were quite senior, their training and expertise were naval, not military. The weaknesses were not only of personnel. While the navy provided medical support in terms of personnel, it did not initially provide any medical supplies, and even when it did, it could not provide the field ambulances, the supply train, the military administrative and quartermaster staff, and (vitally) the divisional artillery; however, as will be seen, it came up with some imaginative substitutes. Then, while the division was still forming up, the Marine Brigade, which was the only part of the division with any military expertise, was deployed on 26 August to spend four days digging trenches near Ostend before returning to Britain. They were then sent out again on 19 September, this time to Dunkirk.

On 2 October Churchill was planning to visit Dunkirk to see a Royal Naval Air Service (RNAS) detachment that had deployed there. While he was en route by train to Dover, he was recalled to a meeting with, among others, Field Marshal Kitchener and Sir Edward Grey, the Foreign Secretary. They were told that the Belgian government intended to evacuate Antwerp. The concern was that if it were to be occupied by the Germans, it would put the Channel ports used to supply the British Expeditionary Force at risk. Churchill offered to visit and report on the situation. With Cabinet agreement to dispatch the RND if necessary, Churchill then went to Antwerp and drove around the defences in an open Rolls-Royce. Presumably this was one of the vehicles being modified by the RNAS as armoured cars.[4] Meanwhile, on 3 October the rest of the division followed the Marine Brigade in crossing the Channel. British responsibility towards Antwerp actually lay with the army, but the British army at this point had only six divisions in France and two more in the United Kingdom. There were no spare formed divisions, trained or otherwise, to be sent to Antwerp. By 5 October Churchill 'had convinced himself that Antwerp's continued resistance depended on his remaining in the city'. Accordingly he sent a telegram to the Prime Minister suggesting that he resign as First Lord to take command of the defence of Antwerp. This caused open laughter at a Cabinet meeting and, despite Kitchener's support for Churchill's suggestion, Asquith ordered Churchill home.[5]

The RND at least now had a commanding general, Major-General Sir Archibald Paris of the RMA, if little by way of equipment or training. Thus the RND went to war, badly equipped and inadequately trained, but, if contemporary accounts are to be believed, in good spirits. Among the officers were three 'temporary acting sub-lieutenants' who were to achieve fame for quite differing reasons: Arthur Asquith, son of the then Prime Minister, who was to finish the war as a brigadier general with a DSO and two bars; Bernard Freyberg, who was to end the war as a brigadier general with a Victoria Cross; and Rupert Brooke, the poet.[6]

The division moved up from Dunkirk to Antwerp by train on 4 October. From the start the division was reminded that it was a naval formation. They were warned that the train might be attacked; if so, they were ordered that Howe Battalion was to deploy from the port side of the train, Anson Battalion from the starboard. Somehow the navy had conjured up an armoured train with four 4.7-inch naval guns to provide artillery support, but as one participant recorded, the German army included the siege guns of up to 15-inch calibre, which had been used to take the fortress of

Fig. 4.2. Major-General Sir Archibald Paris, Royal Marine Artillery.

Liège. The RND dug defensive trenches when they arrived at Antwerp, but they were attacked by the Germans on 6 October in overwhelming force. They withdrew; one of the Royal Marine battalions had to fight its way out, eventually rejoining the division only 150 strong. Thirty-seven officers and 1,442 ratings, comprised mostly of Hawke, Collingwood and Benbow Battalions, felt they had no alternative but to cross the border into the neutral Netherlands, where they were interned. One of the officers, Sub-Lieutenant Grant, was arrested by his commanding officer for refusing to cross into neutral territory. He hung back and waited until his commanding officer had crossed into the Netherlands. He then reasoned that his officer no longer had authority over him and led 35 men back to Allied lines. Of the remainder, 57 had been killed, 138 wounded and 936 made prisoner. The division returned to England. Churchill was the subject of significant adverse press comment; even the Prime Minister in a private letter referred to the affair as a 'shambles'.[7]

When the division returned to England it was reformed. It retained its depot at Crystal Palace, where officer training was carried out. The shortage of officers with infantry skills was compounded by the army's withdrawal of many more Guards officers for army service. The navy had to evolve its own training regimes very rapidly. Normally an officer in training was first taught how to obey orders before learning how to give them. The RND undertook officer training rather differently:

> The atmosphere at the Crystal Palace was rather different; the officer's time of training there was emphatically not a period in which he learnt how to carry out orders, but what orders to give, and how to get them executed; [...] not a period in which he was taught elementary tactics, but rather the underlying principles, knowing which he could not only solve minor tactical problems, but could face [...] the practical business of training troops. [...] If the training of officers was [...] a little ahead of that provided elsewhere in 1914 and 1915, that given to candidates for commissions was actually without parallel till the Cadet battalions were formed nearly two years later, and that given to aspirant N.C.O.s was in advance of anything attempted throughout the war.[8]

The divisional engineers were formed and trained just outside Dover. While the medical officers were from the Royal Navy, medical-branch ratings were largely recruited from the St John's Ambulance. Apparently thought was even given to establishing a divisional cavalry component to be named the Royal Naval Hussars, using as its template the mounted Boer commandos. Despite even getting as far as having a uniform designed, it came to nothing, although a divisional cyclist company was formed, based at Forton Barracks in Gosport. It proved difficult to establish the one remaining component of the division, the artillery, because War Office cooperation was lacking.[9]

The division itself moved to Blandford Camp in Dorset, just outside the market town of Blandford Forum on the edge of Salisbury Plain. Nelson Battalion was the first to move in November 1914, and the rest of the division followed. The move was complete by the end of January 1915. This proved to be fortuitous. Unlike any other British division, it retained its own depot at Crystal Palace and its base at Blandford. These did not come under army jurisdiction, and this allowed the division to build an *esprit de corps* unique in the British land forces. Men went from Blandford

to the division wherever it was deployed and returned for courses, medical
treatment and so on, before again going out to the division.

While he remained First Lord of the Admiralty, Churchill maintained
a close interest in 'his' division, visiting and inspecting it on 17 February
1915. A week later he was present when the King inspected the division
at Blandford. It was inevitable that it would be chosen for Churchill's next
scheme, the attack on the Dardanelles.

The Dardanelles, or Gallipoli Campaign

The Gallipoli Campaign was noted for the appalling losses sustained by
the Australian and New Zealand forces, the Australian and New Zealand
Army Corps (ANZAC), still rightly remembered in those countries to
this day. Often forgotten is the involvement of the Royal Navy – ashore,
in the air and afloat.

The Dardanelles and Gallipoli are used almost interchangeably, but in
general the former is usually taken to mean the maritime aspects of the
campaign and Gallipoli the land campaign on that peninsula. The Straits
of the Dardanelles have always been a strategically important commercial
waterway, linking the Black Sea with the Mediterranean. They run through
Turkish territory and there is a continuous current through it from the
Black Sea into the Mediterranean, which, where the straits narrow, is very
powerful. The importance of the straits was as a supply route to Russia,
an ally of Britain and France, but as Russia's only warm-water access to
Europe, it was also a route for Russian exports, let alone all the traffic that
normally came down the Danube. Before the war 90 per cent of Russia's
grain exports went by sea through this waterway, most of which went,
ironically, to Germany.

Whether Turkey would have entered the war on the German side if
the matter of the Turkish battleships' construction in England had been
handled more sensitively, and if the *Goeben* and the *Breslau* had not successfully arrived in Turkish waters, is a matter of continuing debate. Certainly,
despite a lot of German wooing before the war, Turkey had been friendly
to Britain. At the start of the war Turkey was neutral, and the two German
ships were taken into the Turkish navy. The ships retained their German
crews, but they were inducted into the Turkish navy and even issued with

fezzes in place of their naval headgear! Tensions had been rising, and the Germans had taken over key positions in forts commanding the Dardanelles and even laid minefields in the straits and adjacent waters.[10]

One of Churchill's abiding characteristics throughout his life was a restless aggressiveness. He was not content to attempt to strangle Germany economically by the use of naval power; he wanted to 'do something'. As First Sea Lord, Fisher spent a lot of time and effort trying to curb Churchill's enthusiasms. At the beginning of 1915 Churchill was largely focused on seizing the island of Borkum in the North Sea. However, he became increasingly interested in the Dardanelles. Without the full support of professional advisers in the Admiralty and those in the War Office, he took a scheme to the War Council, effectively the War Cabinet, on 13 January 1915. He proposed a purely naval attack. Ships would bombard Turkish forts and force a passage through to the Sea of Marmara. Such an attack would require few land troops, which would only be needed for raids to demolish forts, and the operation could be broken off at any point if success seemed unlikely. Churchill felt that an Allied fleet arriving off Istanbul would lead to a speedy collapse of Turkey as a belligerent power, hence the initial plan for a purely naval action.[11]

To block any possible foray by the *Goeben* and the *Breslau*, there was already an Allied naval force in the Eastern Mediterranean: two British battlecruisers, two French pre-Dreadnought battleships, two cruisers, 12 destroyers and six submarines. They were under the command of Vice Admiral Carden. He was not held in the highest regard by his colleagues, but with Troubridge at home awaiting court martial after the *Goeben* and *Breslau* affair, Carden was given the command at sea rather than remaining commanding the dockyard in Malta.

The straits had been mined, but a channel through the minefields had been left. On 27 September German sailors were found to be aboard a Turkish destroyer coming out of the Dardanelles, which was a breach of neutrality laws. Carden's squadron turned the destroyer back. Without consulting the Turkish government, a German colonel in command of the forts at the entrance to the straits closed the waterway, darkening the lighthouse essential for navigation and closing the gap in the minefields. Under Admiral Souchon, now nominally of the Turkish navy, the two former German warships bombarded Russian fortifications on the Black Sea coast.

A British ultimatum demanding that the Germans leave the ships was met with a Turkish declaration of war.

Meanwhile, in late January and early February, the Turkish army attacked the Suez Canal. To do so they advanced through Palestine, then part of the Ottoman Empire, taking with them galvanized-steel assault boats for the canal crossing. The main attack came on 4 February, but was beaten off relatively easily by a combination of naval gunfire and three divisions of British and ANZAC troops. The threat to the canal was actually never serious; indeed Kitchener as War Minister appears to have been very relaxed about it from the outset. However, with the threat to the vital canal removed, there were now troops available in the Eastern Mediterranean theatre for other tasks.[12]

Another factor to be considered was the Balkans, or as the area was then termed, the Near East. The war had started because of an Austro-Hungarian ultimatum to Serbia, and there was continuing fighting between the two. On the Eastern Front Russia was being forced back by Germany. Bulgaria was being financially pressured by Germany, and Romania was very exposed but neutral. Greece, Turkey's long-time enemy, was also neutral but very cooperative with the Allies, which included making the island of Lemnos available as an Allied base.

The first phase: the attempt to 'force' the straits

The French had considerable interest in this area of the Mediterranean, and thus, despite misgivings, agreed to participate. Their concerns centred on the lack of any troops to seize the Turkish fortifications after they had been neutralized by naval gunfire. The naval forces now assembled for the bombardment included HMS *Queen Elizabeth*, a brand-new super-Dreadnought, the battlecruiser *Inflexible*, 12 pre-Dreadnoughts, 20 cruisers and destroyers, six submarines, 20 minesweeping trawlers, the seaplane carrier HMS *Ark Royal* and support ships. The French added four pre-Dreadnoughts, seven cruisers and destroyers and four submarines.[13]

Even before the operation started there were reservations in London and Paris. The French suggested delaying naval operations until troops arrived, to no avail. The only troops initially were a division that the French dispatched and two marine battalions from the RND, which had been deployed to undertake minor landings in February. Now the 29th Division and the ANZAC from Egypt were earmarked to move to Lemnos in March.

Fairly soon Admiral Carden's health broke down and he was relieved by Admiral de Robeck. Commencing on 19 February, the bombardment initially went well. It had been planned that the ships would methodically progress through the straits, destroying the fortifications in turn as they did so, but it rapidly became apparent that naval gunfire with its flat trajectory, combined with a lack of forward spotting to control it, made it less effective against shore emplacements than had been hoped. The civilian-manned minesweepers also had a difficult time progressing against the current, which made them much easier targets for the Turkish guns ashore; they were relatively motionless against the land while steaming against the strong current. Churchill therefore ordered a speedier 'advance', accepting the risk of losses. While the shore batteries continued to inflict losses in particular on two French battleships, it was the minefields that were to cause the next attack on 18 March to stall. One French and two British battleships were sunk, and *Inflexible* was badly damaged.

Fig. 4.3. The pre-Dreadnought battleship HMS *Irresistible* sinking after striking a mine.

The landings

Admiral Keyes, Admiral de Robeck's chief of staff, reorganized the mine-sweeping, using more powerful, and therefore faster, destroyers instead of trawlers. De Robeck decided to wait until there were troops ashore, as well as for further naval reinforcements. The delay between attempting to force the straits and the subsequent landings allowed the Turks and the Germans to reinforce the peninsula and, more critically, to resupply their guns with ammunition, which had been running low.

The RND had followed its two Royal Marine battalions from Blandford to Lemnos for the operation. Despite the dreadful losses that were to follow in battle, British (but not Australian) popular history only remembers the loss of one young officer, the poet Sub-Lieutenant Rupert Brooke, who succumbed to illness on the day of the landings.

General Hamilton, who was in command of the land operations, decided to land on five separate beaches. The night before the landings proper, a platoon of Hood Battalion was ordered to create a diversionary landing. Instead, the diversion was successfully created by one man, Lieutenant Commander Freyberg, who swam two miles ashore in the Gulf of Saros with a small raft carrying flares to cause the diversion. He then swam back, despite coming under enemy fire. He was awarded a DSO for this exploit.[14]

The first landings took place on 25 April. The British 29th Division landed on a series of beaches on Cape Helles, the ANZACs on the Gallipoli peninsula. Elements of the RND landed on four of the beaches, but were very split up and did not land as a division. One platoon of D company of Anson Battalion landed with A company on 'V' beach, and the other companies and the battalion headquarters landed on 'W' and 'X' beaches.

There were no specialized landing craft; most troops were ferried ashore in boats rowed by sailors. In some cases these had to be towed two miles or more to the beach by steam pinnaces, although HMS *Euryalus* earned the gratitude of the Lancashire Fusiliers for closing on 'W' beach and braving the Turkish guns in order to shorten the row ashore. Nevertheless, only two boats out of 24 in the first wave made the shore, where the Lancashire Fusiliers would win six Victoria Crosses despite dreadful losses.[15]

The only attempt to provide a specialized landing ship was on 'V' beach, where a modified collier, the *River Clyde*, was intentionally run aground. She

Fig. 4.4. A solid silver model of a naval cutter, presented to HMS *Euryalus* by the Lancashire Fusiliers in gratitude for her support efforts on 25 April 1915.

Fig. 4.5. *River Clyde* ashore on 'V' beach.

had been fitted with limited armour and had holes cut in her side to allow troops to disembark across barges tied alongside the ship to the shore. Due to the heavy fire from the Turkish troops ashore many of the ropes used to secure the barges were cut, and two midshipmen, Drewry and Malleson, each won the Victoria Cross for re-securing the barges under fire.

Unlike army brigades, the RND did not yet have specialized machine-gun companies, so to provide fire support the *River Clyde* had a sandbagged battery of 11 machine guns manned by RNAS personnel under the command of Lieutenant Josiah Wedgwood.

All the landings were attended by heavy casualties. The 'Y' beach landing was initially successful but not reinforced, and following a Turkish counterattack the survivors were taken off. Nowhere did the troops reach the high ground above the beaches that day or subsequently.

At sea the role of the navy was to support the army and the 11,000 men of the RND ashore. They provided both logistic and gunfire support. Their task, however, was to be made much harder by the arrival of German submarines. Meanwhile the Austrians, being aware that Italy might enter

Fig. 4.6. Three midshipmen going ashore for a picnic on Limnos after the landings. Two of them, Drewry and Malleson, had just been awarded the Victoria Cross for their part in the landings.

the war on the side of the Allies, declined to send any of their submarines
from the Adriatic to the Dardanelles. The Germans sent small *UB*-type
submarines overland in sections to Pola in the Adriatic, where they were
reassembled, and thence to the Dardanelles by sea. One was lost en route,
and the other two were actually used in the Black Sea.

U-21 was deployed from Germany to the Mediterranean by sea. Admiral
de Robeck had been forewarned of the submarine's arrival, and instituted
a policy whereby transports would offload at Mudros and the men and
store would be taken to the beaches by lighter and small craft. Despite
early successes – *U-21* sank two pre-Dreadnought battleships, *Triumph* and
Majestic, and a transport, the *Merion*, which had been disguised to look like
the battlecruiser *Tiger* – German submarines thereafter had limited success
against the smaller vessels subsequently used by the Allies.[16]

Meanwhile, the British and French navies had been using submarines in
support of the Dardanelles campaign. The French did not have any success
with their submarines. Not so the British. Even before the campaign proper
opened, Lieutenant Holbrook, in the elderly submarine *B11*, won a Victoria
Cross for sinking a Turkish battleship near the entrance to the Dardanelles.
In the modern *E11* and *E14* respectively, Lieutenant Commanders Nasmith
and Boyle were awarded the Victoria Cross for their exploits. In total the
British submarines not only penetrated as far as Istanbul, sinking another
battleship, but they sank 25 transports and badly damaged a further ten, as
well as various small craft. This did significant damage to the Turks because
the poor quality of the roads onto the peninsula meant that most of their
supplies had to come by sea. *E11* and *E14* also used their deck guns to
bombard the one road there was to Gallipoli in an attempt to prevent troops
and supplies from being brought onto the peninsula by land.[17]

Unfortunately the political will behind the campaign was evaporating.
Fisher was increasingly unhappy with Churchill's management style, which
to Fisher's perceptions intruded on the First Sea Lord's prerogatives. He
had from the start, despite his own misgivings, loyally supported the First
Lord. However, the increasing drain on naval resources, which Fisher saw
as weakening the navy in its major theatre, the North Sea, led him to resign
on 15 May 1915. This cut the political ground from under Churchill's feet.
He had been disliked by the Conservative opposition ever since he had left
that party in 1904, and there was personal antipathy between their leader,

Fig. 4.7. Lieutenant Commander Boyle and the crew of *E14*.

Bonar Law, and Churchill. For Prime Minister Herbert Asquith, the support of the Conservatives was essential to his government's survival. He chose to form a national or coalition government, with Arthur Balfour as First Lord of the Admiralty. Churchill now left the naval scene, until he returned as First Lord in September 1939.

Initially the War Council was minded to continue with the Gallipoli Campaign, and there were further landings at Suvla by five new British divisions. The French planned to land four divisions on the Asiatic side of the straits, but while the Allied lodgement was secure, there was no realistic chance of advancing, certainly not sufficiently to allow the fleet into the Sea of Marmara. On 14 October 1915 Bulgaria entered the war on the side of the Central Powers, which meant that there was an immediate prospect of German land reinforcement for Turkey.

The withdrawal

On 16 October General Hamilton was peremptorily sacked and replaced by General Monro. Almost immediately on taking command he recommended that the peninsula be evacuated. Kitchener himself came out to see the theatre, and after consideration of the alternatives he also recommended that the evacuation take place. It was anticipated that casualties of 30 per cent would be incurred during a withdrawal. In the event, unlike much of the rest of the campaign, the evacuation was a masterpiece of planning and execution. The ANZAC Suvla sector was evacuated on 19 December without casualties, and 17,000 men were evacuated from Cape Helles by 03:45 on 9 January without loss.[18]

Dardanelles or Gallipoli, however named, was little short of a disaster, and at the time it seemed to end Churchill's political career. In 1940 Churchill was to show dogged, single-minded determination to overcome all obstacles, obstructions and objections, many of them originating among erstwhile allies. He showed many of the same characteristics in 1915. Unfortunately in 1915 he either overruled or ignored professional advice and tended to hear what he wanted to hear. His reputation was to be severely damaged for many years.

The Royal Naval Division in France

After its withdrawal from the Dardanelles, the RND remained in the eastern Mediterranean while its future was debated. For a time one brigade held part of the line in the Salonika theatre. This was an Allied lodgement intended to counter the Bulgarians, but it was to be relatively inactive until 1918. Eventually the decision was made that the division would go to France.

Once it arrived on the Western Front it went to a back area around Abbeville for reorganization and training. In June 1916 the division became the 63rd Royal Naval Division, with the two Royal Navy/Royal Marine brigades being numbered the 188th and 189th, and came under the Army Act for discipline. Until then it had no artillery and, most importantly, limited numbers of machine guns. In the army, there was a separate Machine Gun Corps (MGC) that provided units to army battalions, the Vickers machine gun being a brigade rather than a battalion weapon. Battalions were issued with Lewis guns. The RND now gained its own artillery, machine-gun companies and support arms and was augmented by an army brigade, the 190th.[19]

Fig. 4.8. Ratings and marines of the Royal Naval Division
behind the lines in France, wearing beards and naval caps.

Uniquely among British divisions, the RND retained its identity.
Units tended to come and go from army divisions, and in any case army
personnel historically related to their regiment, not to their brigade or
division, while in the RND officers and ratings looked to the division
and the service. Internal discipline and pay remained naval, as did service traditions, so officers and ratings could and indeed did grow beards,
which were forbidden to the army. They were also careful to point out,
if necessary, that as the senior service they took precedence over army
formations on parade.

Once considered to be trained and ready for the static trench warfare
in France, the division, as part of IV Corps, moved up to the front under
General Paris's RMLI on 17 July 1916, taking over the Angres-Souchez
sector close by Vimy Ridge. The Battle of the Somme was raging, but the
division was in a quiet sector and little of note happened until October,
when it was transferred to V Corps, part of the Fifth Army. General Paris
was severely wounded on 14 October while visiting one of his brigades
in their trenches. This was a severe shock to the whole division; far more
than army units, they looked to their divisional commander as an individual. There was a significant drop in morale, particularly as the army
Major-General Shute, who took over, tried to change its naval practices

and customs, even including their salute (the navy salutes with palm of the hand downward, the army with the palm forward). His reports made to higher formations were contradictory, but he obviously had no regard for the navy or its ways. Unfortunately for him, one of the divisional officers, Lieutenant A.P. Herbert of the RNVR, had aspirations to being a poet. One poem, concerning an officer being ordered to remove his beard, was published during the war in the humorous magazine *Punch* as 'The Ballad of Godson's Beard'. A far more scatological poem, which circulated throughout the division with regard to the general's concerns over field hygiene, concluded:

But a shit would be shot without mourners
If someone shot that shit Shute.[20]

General Shute had assumed command just before the division took part in the battle of the Ancre, part of the Somme battle. The division attacked on 13 November. Commander Asquith of Hood Battalion had been wounded during the preparation for the battle and his place had been taken by Lieutenant Colonel Freyberg, who had won a DSO in the Dardanelles and seems to have adopted army rank. The division captured all of its objectives, not common during the Battle of the Somme. Freyberg won a Victoria Cross during the capture of the fortified village of Beaucourt. He was wounded and command devolved on Lieutenant Commander Egerton, DSO. Unfortunately the division suffered serious losses, with 100 officers and 1,600 ratings and marines killed, and 160 officers and 2,337 ratings and marines wounded.

Despite its performance, or because of it, General Shute now moved to fill empty slots with 'proper army' officers, stating that 'the gallantry of the RND on Ancre only emphasizes the need to re-organise'. The future of the RND was now fought at the highest level. The army wished to transfer the naval and marine personnel of the division to the army. The Admiralty firmly resisted. At the meeting held on 5 February 1917 chaired by Lord Derby, the Admiralty in the person of the First Lord himself, now Sir Edward Carson, very firmly supported the division remaining naval and also drew attention to General Shute's personal antipathy to the Royal Navy. On 17 February General Shute was relieved by another army general,

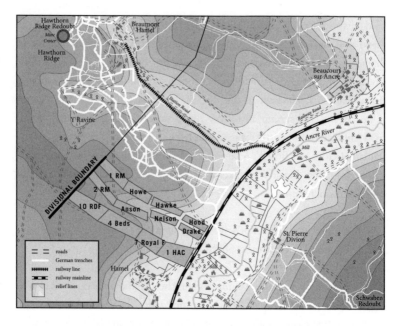

Fig. 4.9. Map showing the deployment of the Royal Naval
Division at the opening of the battle of the Ancre.

Major-General Lawrie, who was a much different personality and far more
sympathetic to the navy and its ways. He was to command the RND until
nearly the end of the war. The army formally withdrew its proposal and
the RND remained part of the navy for the remainder of its existence.

After the battle of the Ancre, the RND was rested until January 1917.
During this period it was brought back up to strength before going back
into the same sector of the front. In February a successful assault was carried out on the German lines, achieving its objective, but at the cost of 24
officers and 647 petty officers and men, with a high proportion of those
killed. This was followed by a further attack on 17 February, fortunately with
lighter casualties. The resulting German counterattack was 'annihilated',
but in further fighting the marines lost over 400 officers and men before
the RND was relieved at the beginning of March. After a rest period, the
division again moved up to the Arras area. They were due to attack a week
before the Nivelle Offensive by the French army, in which much political
hope was invested. Unfortunately, the Germans anticipated the offensive
and pulled back; in essence, the French punch was to hit empty air.

Following the failure of the attack, the French army mutinied. The British took the offensive at Passchendaele largely to distract the Germans from their weakened ally. While the battle itself was unsuccessful, the RND in capturing Gavrelle 'indicated the tactical proficiency of some units of the [British Expeditionary Force]'. During this dreadful battle the division suffered heavy losses. To give a measure of the casualties during this period, between November 1917 and opening of the German attacks in 1918 the RND lost five battalion commanders and was to lose four more during the 1918 retreats. The loss in junior officers was proportionately greater, but it was noticeable that the quality of the junior officers coming forward remained high. It was about this time that a senior army officer asked an admiral why 'naval officers seemed to be much more self-reliant than their army brethren'.[21]

The German 1918 offensives

The defeat of Russia in 1917 and the Treaty of Brest-Litovsk freed up a lot of German troops for employment on the Western Front. The German High Command staked everything on a series of major offensives in early spring 1918, intending to defeat the Allies before the United States army could play a significant part.

In the preliminary bombardment with mustard-gas shells, which started on 8 March, Hawke and Drake Battalions lost 36 officers and 990 ratings. The 189th Brigade headquarters lost the brigade commander, the brigade major and all the clerical and signal staff – this just before the biggest German offensive since 1914. The attack began on 21 March and while the RND held, loss of support on the flank meant that they had to retreat, and again on 23 March. The historian of the division records: 'Never did the men of the Division show more clearly how superior was the quality of their discipline, than when the 189th Brigade were operating round High Wood on the evening of March 24th.' The brigade, along with some other divisions, 'saved the British Army; where these divisions were, there were no gaps'. The Division even launched decisive counterattacks on 26–7 March, despite having lost around 6,000 men and having received no reinforcements. The average strength of the battalions, normally between 800 and 1,000 men, was reduced to 250, but they held.[22]

When they were eventually relieved, they received replacements for their casualties and undertook training before again going up into the line with each battalion at a strength of 25 officers and 800 men. The reinforcements were men new to France. It is a tribute to them all that the division was able to resume attacking raids on 18 May, having absorbed this number of men; however, casualties continued. A raid on 24 May cost 18 officers and 210 men, although a further night-time raid on 12 July carried out by Drake Battalion cost the enemy 44 casualties for one of their own.

Thereafter, the Allies were on the offensive. After the preliminary battle of Hamel came the great victory at the battle of Amiens. Now largely forgotten, this was, in the words of General Ludendorf, 'a black day for the German Army'. In the advance that followed, the RND led XVII Corps's attacks. Despite German counterattacks (using captured British tanks) the RND achieved major advances, taking many German prisoners. The RND continued as part of the spearhead of the advance, including the crossing of the Canal du Nord and penetrating the Hindenburg Line. Probably the RND's finest feat came during the battle for Cambrai (the scene of an earlier battle). The division played a large part in cutting off the retreat of the German left wing, advancing more than seven miles in four days, which included attacking prepared German positions on four occasions. During this period Commander Beak of the RNVR, commanding Drake Battalion in an attack from the very front, won a Victoria Cross (to go with a previously won DSO and MC).[23]

After the armistice, the RND was rapidly disbanded. While it was one of many infantry divisions that fought in World War I, its structure and ethos marked it out as being very different from the others. After an inauspicious start in Antwerp, it could fairly claim to be one of the better Allied fighting divisions, probably due in large part to the original way the navy trained its infantry officers.

The Zeebrugge raid

Elsewhere on the Western Front, Antwerp and the other ports continued to be of concern to the navy. The Germans used the ports as submarine bases to attack both military and civilian shipping. The journeys from the bases at Zeebrugge and Ostend to the areas where such targets could be found were short, meaning that the German navy could use its shorter-ranged coastal U-boats.[24]

Many efforts had been made to disrupt the German bases, to little avail. The RNAS had attacked them, and from August 1915 they were subject to bombardment by monitors. These were small flat-bottomed ships mounting one or two guns, typically a pair of 12-inch, 14-inch or 15-inch calibre.[25] Ultimately there were two monitors, the *General Wolfe* and the *Lord Clive*, which carried a single 18-inch gun. The two problems for the monitors were that firstly, while the guns were extremely accurate, controlling them to hit the target was problematic – as it had been in the Dardanelles – and secondly, of course, the Germans, who fired back. The latter meant that bombardments had to be carried out from some distance, making the first problem even more difficult. Spotting from aircraft was employed, as were portable tripods to form temporary islands for spotters to observe the fall of shot. In foggy weather taut-wire measuring gear was used. 22-gauge wire was paid out from the stern of a ship; this allowed the distance steamed to be measured accurately (the error was 0.2 per cent of the distance run), and by knowing the direction steamed an accurate ship's position could be determined to allow firing in fog and mist. However, despite such increasingly ingenious ideas, including torpedoing the gates and the planned 'great landing' of an artillery gun to bombard the gates from shore, the lock-gates and canal at Zeebrugge and Ostend were not damaged. U-boats could still get to sea. Increasing submarine activity and, with the advent of unrestricted submarine warfare in 1917, merchant-ship losses forced the decision to launch an amphibious attack.[26]

The basic plan was to block the canal entrances at Ostend and Zeebrugge with blockships, old cruisers packed with concrete and explosives which would be sunk in the entrances to the locks. This would prevent German submarines and destroyers from leaving Bruges and getting to sea. At Zeebrugge there was a complication: the entrance to the canal was protected by a mole (see Fig. 4.11) on which there were numerous gun emplacements that would be able to fire on the blockships as they approached. The mole was connected to the shore by a 300-metre viaduct; thus at Zeebrugge it was necessary to storm the mole and simultaneously to prevent reinforcements getting to it by destroying the viaduct. This was to be achieved by positioning two old *C*-class submarines with explosive charges to blow it up, while at the same time an old cruiser, the *Vindictive*, would be laid alongside the mole and Royal Marine and naval landing parties from her would capture the guns on the mole. *Vindictive* was modified with extra armour, an 11.5-inch howitzer, flame-throwers, mortars, machine guns and, all importantly, boarding ramps for the landing parties to disembark.[27]

In overall command was Rear Admiral Keyes, who had been Admiral de Robeck's chief of staff in the Dardanelles. He was an inspiring leader but was impetuous, and has been described by a contemporary, Admiral Richmond, as having 'courage and independence [...] though very little brains'. He now commanded the Dover Patrol, which was to launch the operation, but had just come from the Admiralty, where as director of plans he had been responsible for the inception of the operation. The detail-planning had been undertaken by his predecessor at the Dover Patrol, Rear Admiral Bacon. The first attempt to launch the attack, known as Operation *ZO*, came in early April 1918. Unfortunately two attempts at launching the operation were aborted and, after the second, HM Coastal Motor Boat 33 was captured. Thus the Germans knew something was planned; then Keyes, instead of waiting for a moonless night and a high tide, which at first he had insisted were essential, decided to attack at the earliest midnight high tide, the night of 22–3 April. This was a moonlit night. To make matters worse, during the approach the wind changed and blew away a smokescreen that had been laid as concealment. *Vindictive* got alongside the mole, but disembarkation was delayed by casualties and damage sustained during the approach. An intensive battle at close range, at times hand to hand, was now fought on the mole.[28]

The submarine attack on the viaduct was entirely successful and prevented German reinforcements getting onto the mole. The first of the blockships, *Thetis*, received the undivided attention of the guns on shore from ranges as little as 100 yards. Because of damage depriving her of her engines, her commanding officer scuttled her; however, she diverted the shore gunners' attention from the second ship, *Intrepid*, which for reasons never fully explained was scuttled short of the lock-gates at the entrance to the canal. *Iphigenia* was similarly damaged and scuttled short of the entrance.[29]

At Ostend, forewarned by information gleaned from HM Coastal

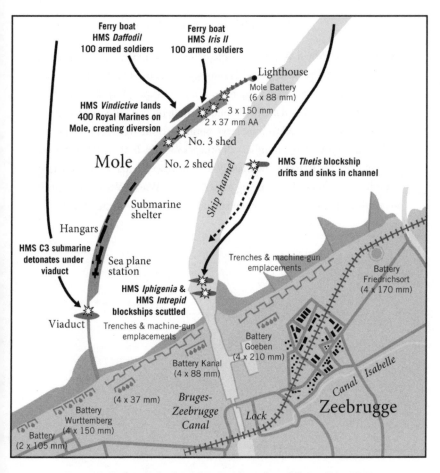

Fig. 4.10. Zeebrugge harbour showing the mole and the viaduct. HMS
Vindictive's position and the final positions of the three blockships are shown.

Motor Boat 33, the Germans moved the navigation buoys. Relying on
them to get into their planned positions, the blockships were scuttled in
the wrong places.

The whole operation lasted barely an hour, during which there were
600 British casualties. No less than eight Victoria Crosses were awarded,
but the canals were only blocked for a few days and submarines were once
again able to exit into the North Sea. However, the raid was a major fillip
to national morale.

The amphibious Royal Navy

For the navy, World War I was a major learning experience. It had not conducted large-scale opposed landings such as the Dardanelles before. It had had to evacuate an army from a hostile shore, but never quite in the manner of the Dardanelles. While it had deployed sailors and marines ashore, it had never had to train large numbers of men for such a role, nor had it had to train large numbers of officers rapidly. In World War II it was to fight from the first day and manage an unforeseen expansion to the largest it had ever been. The training experience of World War I was to prove invaluable; it was, after all, undertaken by many of the same officers. [30]

CHAPTER 5

The Battle of Jutland

The battle of that great day... a place called Armageddon.

BR 401 (The Holy Bible)
Revelation 16:14–16

Ever since the outbreak of war the entire Royal Navy, not just the Grand Fleet, had been waiting for the *Hochseeflotte* to 'come out'. Initially the British had expected the Germans either to support their army's advance into Belgium and France, or even to invade the United Kingdom. At the same time, the Germans expected the British to come into the German Bight and institute a close blockade. Neither did, or had planned to do, what the other expected. Alien to the British way of thinking, the Germans, particularly the Kaiser, wished to preserve their battlefleet rather than use it, to retain a 'fleet in being' as an existential threat. The Kaiser repeatedly gave instructions that the fleet was not to engage an enemy unless they had overwhelming superiority. German strategy was therefore aimed at wearing down British numerical superiority in Dreadnought battleships until equality in numbers was achieved. To do so they aimed to catch an isolated segment of the Grand Fleet so that it could be destroyed piecemeal. The various bombardments of the east-coast towns had been part of that strategy and had led to the battle of Dogger Bank.

The British were increasingly frustrated by the Germans' refusal to fight. Many schemes to attack them in Wilhelmshaven were studied and discarded, but as Jellicoe was to conclude in January 1916: 'Until the High Seas Fleet emerges from its defences, I regret to say I do not see that any offensive action is possible.'[1]

Reinhard Scheer, newly appointed as commander in chief of the *Hochseeflotte* in early January 1916, saw an opportunity. Because the US government had threatened to break off diplomatic relations, the German government instructed him to cease the unrestricted U-boat campaign against Allied and neutral shipping,[2] and so he withdrew all his U-boats from the shipping lanes. He now had at his disposal a large number of submarines and conceived a plan that, in essence, aimed to draw Beatty's battlecruisers over a submarine trap. Initially he planned to bombard Sunderland on the British north-east coast, but modified this to cruising off Norway in a manner that threatened British trade between the two countries. The plan was that once a portion of the Grand Fleet, most likely Beatty's Battlecruiser Fleet, had been weakened by the submarines, firstly Hipper's battlecruisers would attack them, followed by the *Hochseeflotte* before the mass of the Grand Fleet could intervene. Use of Zeppelins for reconnaissance was an integral part of Scheer's planning. This required good weather, both so that they could fly and so that they could see what was going on. He particularly wanted them to warn him of the approach of superior forces.

Changes in the organization of the Grand Fleet

By May 1916 there had been some organizational changes following the attack on Lowestoft in March by German battlecruisers temporarily under the command of Admiral Bödicker. To reassure the public, the 3rd Battle Squadron (3BS), the 'Wobbly Eight' and HMS *Dreadnought* were moved south to be based in the Thames.

The Battlecruiser Squadron was now designated the Battlecruiser Fleet (BCF) and was divided into three Battlecruiser Squadrons (BCS). However, for Jellicoe the standard of battlecruiser shooting was of greater concern than the German raids. In particular *Lion* and *Tiger*'s gunnery was poor. The cause was simple: lack of practice. The ships of the main body of the Grand Fleet had ready access to gunnery ranges around Scapa Flow; Beatty's ships based in the Firth of Forth did not. All that was readily available to his ships at Rosyth were ranges for sub-calibre shooting. In a sub-calibre shoot, the guns had a smaller barrel inserted inside them so that they would fire a lighter, smaller shell, which was hardly ideal to train and exercise either the guns' or the directors' crews. To give the battlecruisers an opportunity for full-calibre firing, they were to be sent north to the ranges around Scapa

Flow. Thus it was planned that while the 3rd Battlecruiser Squadron (3BCS) under Rear Admiral Hood was detached, in temporary substitution the 5th Battle Squadron (5BS) would be subordinated to Admiral Beatty and moved from Scapa Flow to the Forth. This was made up of five ships of the *Queen Elizabeth*-class, the first super-Dreadnoughts. They were oil- rather than coal-fuelled and much faster than the rest of the Grand Fleet; they were only a knot or two slower that the swiftest of the so-called 'big cats', the battlecruisers. They were well armoured and armed with the new 15-inch gun, developed in secret as the 'experimental 14-inch'. This was the best British naval big gun until the end of the battleship era.

The purpose for which battlecruisers were designed and how they should be employed has been much debated without a definite conclusion, even to the present day. Jellicoe's 'Grand Fleet Battle Orders' were unequivocal in considering them part of his scouting force. Beatty had a much different view, considering his ships to be part of the battle line. He had been agitating for some time to be given 5BS under his command for that purpose. Beatty's deputy, Rear Admiral Hood, felt that attaching the 5BS was 'a great mistake'. He felt that it gave Beatty the justification to consider his battlecruisers a match for the German battlefleet.[3]

Thus it was that 3BCS under Admiral Hood left the Firth of Forth to go up to Scapa Flow, while 5BS under Rear Admiral Hugh Evan-Thomas arrived on 22 May 1916 and moored in the Firth of Forth. Between that date and when the ships sailed on 30 May for what turned out to be the battle of Jutland there is no evidence that Beatty and Evan-Thomas met, despite the ships' being moored within sight of each other. Since Beatty was to complain after the battle that he had not been able to have 5BS to train with the BCF, it seems curious that he did not make an effort at least to talk to its admiral; after all, this was Beatty who, writing to Tyrwhitt a few weeks earlier, had said that 'one hour's conversation is worth volumes of correspondence'.[4]

Gunnery in the Battlecruiser Fleet

To attempt to make up for their poor shooting, BCF ships had cut corners to fire faster, substituting volume for accuracy. This meant taking major risks. British big guns were termed 'BL'; while this actually stood for 'breech-loading', harking back to the days of muzzle-loading guns, the term was used to designate guns whose ammunition came in two parts, a shell (the projectile) and a separate propellant charge, rather than a single piece of ammunition as would be used by a 'quick-firing', or 'QF', gun. The two parts were stored separately in the ship: the former in the shell room and the latter in the magazine. The propellant used by the Royal Navy was cordite, a mixture of nitroglycerin and nitrocellulose contained in silk bags. It is actually quite difficult to make cordite explode; normally it burns with extreme rapidity, but it will explode if exposed to a flash or an explosion. Hence, in each bag was a charge of gunpowder to set it off when required once it was safely in the gun. The bags were of silk because it burnt when the gun fired and did not leave any residue, which otherwise could eventually clog the barrel.

Fig. 5.1. Clarkson cases recovered from the wreck of HMS *Audacious*.
As can be seen from the damaged case shown, they were thin-walled
and made of brass, which does not generate a spark if struck.

The charges were stored and moved outside the magazine in tubular 'Clarkson cases' which were supposed to be kept closed as long as possible, specifically to protect them from the flash of an explosion or an inadvertent spark. Each case contained two one-quarter charges; thus two cases were required to fire each gun. The charges in their cases were moved up from the magazine on a mechanical hoist to the handling room below the gun turret, where they were removed from the case by the handling-room crew. The charge would pass through anti-flash doors, which would open and close as the charge went up the hoist. These were intended to prevent an intense flash from an explosion in the turret or handling room going down to the magazine. From the handling room, one shell and one complete charge for each gun went up on separate hoists through more anti-flash doors to the turret, where they would be rammed into the gun. The Clarkson case was supposed to be returned to the magazine on the hoist as it returned downwards.

In the Battlecruiser Fleet, to speed up this process so that the guns could fire more rapidly, there came to be a practice of storing both full and empty Clarkson cases in the handling room and even in the turret itself, sometimes with the lids removed. At the battle of Dogger Bank:

> the magazine crew full of enthusiasm, and determined that the guns should not have to wait for cordite, had removed practically every lid from all the cases, piling up the handling room with charges. In addition they filled the narrow passage in the four magazines [one for each turret] with more charges.

Worse, in some ships the anti-flash doors were wired open or even removed. Thus the flash from an explosion in the turret could travel down the hoist, through the handling room and into the magazine where there would be exposed charges.[5]

The efforts of the newly appointed warrant gunner in *Lion*, Alexander Grant, in the short period before the battle made her a much safer ship, and probably saved her from blowing up like three other battlecruisers at Jutland. He restored regulation ammunition-handling practices and, ironically, demonstrated to her captain's satisfaction that doing so would not slow her rate of fire. However, the other ships of the BCF continued with dangerously poor ammunition-handling practices, with the best of intentions, for a faster rate of fire.

Prelude to battle

The Royal Navy's mastery of the new art of signals intelligence now showed its worth. Since the middle of May, Room 40 had been reading signals indicating that the Germans were planning a sortie. First it read the signals dispatching Scheer's submarines to sea, and then it read the daily reports that the submarines made while at sea. This culminated on 30 May when a German signal which read '31G.G. 2490' was intercepted. While the signal was itself not understood (although did the '31' refer to a date?), the fact that it was sent to all units of the *Hochseeflotte* led the Admiralty to order the Grand Fleet to sea. Unfortunately, due to a signalling error, Jellicoe's aircraft carrier HMS *Campania* was left behind, depriving him of air reconnaissance. However, Beatty was accompanied by the HMS *Engadine*, a seaplane carrier.

Thus the Grand Fleet actually got to sea before the *Hochseeflotte* and was able to move to intercept them with overwhelming force. When they did get to sea, the Germans had 21 Dreadnought battleships and battle-cruisers against 37 British. The Germans also took with them the 2nd Squadron, comprised of six slower pre-Dreadnoughts under Rear Admiral Mauve. They were known to their compatriots as the 'five-minute ships'; an estimate of how long they were anticipated to last in a fight with the Royal Navy.

The BCF was steaming south-eastward with the battlecruisers of the 2nd and 3rd Battlecruiser Squadrons in two columns of three, but with Evan-Thomas and 5BS astern and to port, five miles from Beatty's flag-ship. This was an odd place at best to put his most powerful squadron; it has been called Beatty's 'fatal error'. He also had accompanying him three squadrons of light cruisers, including HMS *Galatea*. At about 14:00 Beatty ordered the planned turn to the north to join the Grand Fleet when *Galatea* came on a Danish merchant ship that appeared to be blowing off steam, indicating that she had come to an unexpected halt. *Galatea* increased speed, closed the merchantman and then saw beyond it some German destroy-ers followed by cruisers. She reported by wireless that she was in contact

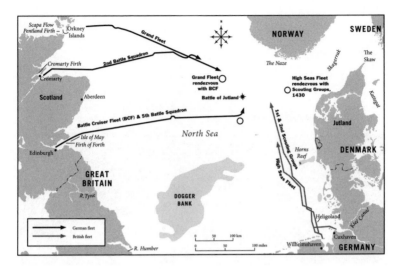

Fig. 5.2. The initial moves by the two opposing forces.
At this point it had been planned for Beatty to turn north
in order to join the rest of the Grand Fleet.

with the enemy. Her initial signal was followed by another, reporting the possibility of larger ships. Beatty ordered *Engadine* to fly off a seaplane. Unlike *Campania*, which had a short flight deck, *Engadine* could only launch her planes by lowering them into the water by crane for them to take off from there. The seaplane took off within 25 minutes. The observer in the plane saw and reported by wireless German light cruisers, but to no avail. While the messages did reach *Engadine*, she was unsuccessful in passing them on to Beatty. The failure of *Campania* to sail with the Grand Fleet and the weather that precluded the German Zeppelins' useful participation in the battle that was to follow meant that there was no other involvement by aircraft from either side until the fighting ceased.[6]

At this point, unknown to the Germans, the Grand Fleet was heading south and, having received *Galatea*'s report, increasing speed. Owing to an error by the Admiralty, Jellicoe had been informed that the *Hochseeflotte* itself had not sailed, so at this point each battlefleet was unaware of the presence of the other. However, HMS *St Vincent*, one of Jellicoe's battleships, informed Jellicoe at 14:28 that she was intercepting strong German wireless traffic on the *Hochseeflotte* wavelength. Jellicoe ordered a further increase in speed.

Fig. 5.3. The Grand Fleet at sea. Note the amount of
smoke that coal-fired ships made. In battle this would add
to the smoke from the guns to affect visibility.

At about the same time Hipper, in command of the German battlecruis-
ers *Lützow*, *Derfflinger*, *Seydlitz*, *Moltke* and *Von Der Tann*, sighted the British
battlecruisers. Beatty had anticipated their likely position and turned east,
hoping to interpose his ships between the Germans and Germany. Hipper
now tried to draw the BCF onto the *Hochseeflotte*, then some 50 miles to
the south and steaming north.

The 'run to the south'

What followed is known now as 'the run to the south'. For some reason,
not understood at the time or later, Beatty delayed opening fire, possibly
because of the time taken in sending a sighting report to Jellicoe.[7]

Beatty knew that the guns of his battlecruisers outranged those of the
Germans; however, because of the delay, the two opposing lines of battlecruisers opened fire almost simultaneously at about 15:45. Unfortunately
there was, just as at Dogger Bank, a problem with fire distribution. This
time *Derfflinger* was left unengaged; thus she could shoot without the
distraction of being shot at.

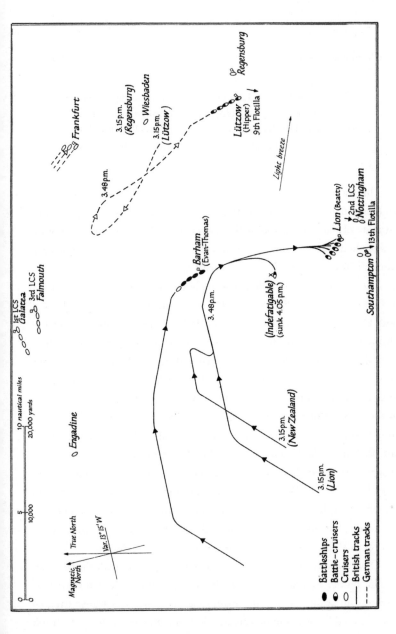

Fig. 5.4. The start of the 'run to the south' at about 14:10. The 5th Battle Squadron, headed by HMS *Barham*, is stationed off port of and five miles astern of the rest of the Battlecruiser Fleet.

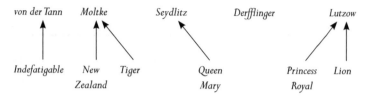

Fig. 5.5. Battlecruiser Fleet initial fire distribution during the
'run to the south'; that is, which ships were firing at which. It can
be seen that HMS *Tiger* and HMS *Queen Mary* initially engaged the
wrong ships, leaving SMS *Derfflinger* unengaged for 10 minutes.

This was compounded initially by the BCF's poor shooting. They took a
lot longer than the Germans to 'find the range', that is correcting the close
estimate of the range given by the rangefinders. This was done by noting
where shells actually fell. Once they did, *Derfflinger*, *Lützow* and *Seydlitz*
were hit. Then *Lion* was hit on 'Q' turret amidships.

The hit on the turret killed or wounded most of the men in the turret,
including the turret officer, Major Harvey of the Royal Marines, who lost
both his legs. The initial explosion was followed by a second as a cordite
charge fell back out of the open breech of one of the guns. Fortunately,
a combination of the reimposed magazine discipline by Warrant Gunner
Grant and the bravery of the turret officer, who flooded the magazine,
meant that the resulting explosion was only of the one additional charge
in the hoist. While the explosion was sufficient to blow off the turret
roof, it did not lead to a magazine explosion and the loss of the ship. If the
anti-flash doors had been open and the handling room full of uncovered
charges, *Lion* would have gone the way that three other battlecruisers were
to go later that day. Nonetheless, the entire turret crew died, either from
the initial hit, by drowning in the flooded magazine or in the secondary
explosion. For his devotion to duty, despite his mortal wounds, Major
Harvey posthumously won the first Victoria Cross of the day.

Then *Indefatigable*, which had been conducting a duel with *Von Der Tann*,
was hit on her forward turret. She blew up, leaving only two survivors.
Lion was hit again. She lost her main long-range wireless transmitter, so
now she could only communicate by wireless with ships close by. At 16:08
5BS came into range, opening fire at 20,000 yards. They quickly found

Fig. 5.6. HMS *Lion* at the moment of being hit on 'Q' turret. The explosion can be seen as the grey column of smoke going vertically up between her funnels through the black funnel smoke.

Fig. 5.7. The 5th Battle Squadron.

the range and started to hit the German battlecruisers, first *Von Der Tann* and then *Moltke*.

Queen Mary was recognized as the best gunnery ship in the BCF, but according to the gunnery officer of the *Derfflinger*, while she was 'shooting superbly [...] As a rule all eight shots fell together. But they were almost always over or short.' She was on the receiving end of the fire from *Derfflinger* and *Seydlitz*; at 16:26 she was hit and a magazine exploded. She rapidly sank with 1,266 men, leaving only two survivors. *Tiger*, moving at 25 knots, was only 500 yards astern of her and had to manoeuvre violently to avoid hitting the wreckage.

One of the legends of Jutland arose out of what followed. *Princess Royal* was then straddled by a German salvo (shells falling both over and short) and observers on *Lion*'s bridge thought she, too, had blown up. At this point Beatty made his famous remark that 'there seems to be something wrong with our bloody ships today' and apparently ordered that the ships alter two points (about 25°) to port, that is towards the enemy. However, there is no record of the signal or the turn; it is all part of the Beatty legend. Beatty did order his destroyers to launch a torpedo attack, which Hipper met with a cruiser and destroyer attack of his own. The exchange cost the Germans a torpedo hit on *Seydlitz* and two destroyers, and two British destroyers, HMS *Nestor* and HMS *Nomad*, were hit in their turn and later sank. Commander Bingham, commanding officer of the *Nestor*, closed the enemy battle line to within 3,500 yards before turning to fire his torpedoes. He was awarded a Victoria Cross for his conduct during this attack. Among the other commanding officers to be decorated for this part of the action was Lieutenant Commander Tovey of HMS *Onslow*, who received the DSO. He, as Admiral Tovey, was to sink the *Bismarck* in World War II, ultimately by torpedo.[8]

Hipper now sighted the *Hochseeflotte* and turned north. At 16:35 Commodore Goodenough, commanding the 2nd Light Cruiser Squadron (2LCS) from *Southampton*, sighted the German battlefleet. He sent a signal reporting this to both Jellicoe and Beatty. He closed the Germans to ensure good visibility and followed up his original signal with one giving the course, composition and disposition of the enemy, exactly what his superiors expected of a scouting cruiser squadron.

The 'run to the north'

Beatty now ordered a turn 16 points (i.e. 180°) to starboard in succession; his ships would turn one after the other. The signal would have been made with three flags: 'Compass Pendant, One, Six'. Thus the order of the ships would remain the same, with the flagship *Lion* leading. The signal was hauled down at 16:41, executing the order. The phase of the battle that followed is known as the 'run to the north'.

Unfortunately 5BS did not receive the signal. *Tiger* had been instructed at the outset to pass Beatty's signals to 5BS by light in view of the distance, but she did not. Thus Beatty, heading north, passed Evan-Thomas, heading south, at a combined closing speed of 50 knots, close to 60 miles per hour. The signal to turn was again made at 16:48, but was not executed for six minutes, leaving 5BS nearly five miles closer to the enemy than they need have been. Obediently, the ships of 5BS turned one after the other as they each reached the same spot in the ocean. Not surprisingly, the Germans fired at the turning point, and *Malaya*, the last ship in line, was the focus of several German battleships' fire. Fortunately her captain had the sense to turn early, breaking formation, and upsetting the Germans' aim. An unanswered question is why Beatty did not order 'Blue Pendant, One, Six', which would have meant that all the ships would have turned at the same time and would have been that much further from the enemy.

During the run to the north the battlecruisers were not engaged, but 5BS was engaging not only the German battlecruisers, but also the leading ships of the *Hochseeflotte*. *Seydlitz* was close to sinking; *Von Der Tann* had no working main armament guns, but kept her place in the line so that British fire would not be concentrated on the other ships. Now, in a reprise of the battle of Coronel, the British line had the setting sun behind them, but it was to be a lot longer before it would actually set. According to Admiral Hipper, the sun combined with the effect of gun and funnel smoke meant that the German 'gunnery officers could find no target although we made a superb one ourselves'. Despite this, the Germans succeeded in hitting the ships of 5BS, while the battlecruisers were now out of range. Throughout the run to the north, Goodenough and the light cruisers *Southampton*, *Birmingham*, *Nottingham* and *Dublin* remained astern of 5BS and in contact with the German fleet, making a string of reports. By judicious alterations

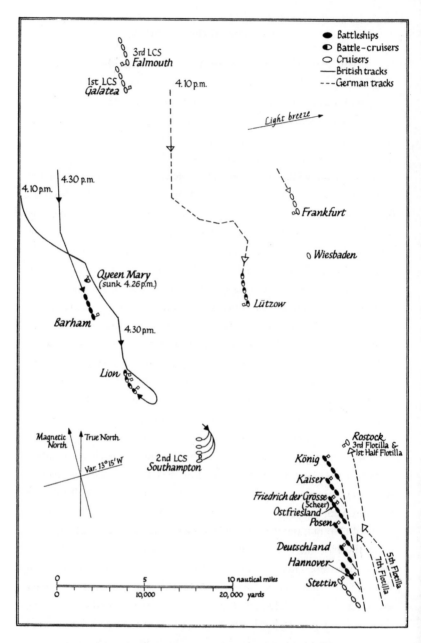

Fig. 5.8. At about 16:50 the Battlecruiser Fleet and its light cruisers encountered the *Hochseeflotte* and turned to the north, marking the start of the 'run to the north'.

Plate 1. *The Famous 5th Battle Squadron* (detail).

Plate 2. *Destroyers Engaging German Cruiser.*

Plate 3. A contemporary German postcard of
Kapitänleutnant Weddigen, the commanding officer of
U-9, which sank HMS *Hogue*, *Aboukir* and *Cressy*.

Fig. 5.9. HMS *Birmingham* under fire.

of course they avoided being hit, despite being as close to the enemy as 13,000 yards.[9]

Rear Admiral Harper was very critical of Beatty's handling of the opening stages of the battle, in particular the positioning of 5BS and the poor shooting of the BCF. He concluded:

> a British squadron, greatly superior in numbers and gun-power, not only failed to defeat a weaker enemy who made no effort to avoid action, but, in the space of 50 minutes, suffered what can only be described as a partial defeat.[10]

The clash of the battlefleets

As soon as Jellicoe received Beatty's initial report that he was in action, he detached 3BCS to use their superior speed and to head south to support Beatty. Commanded by Rear Admiral Hood, 3BCS was made up of the three oldest battlecruisers: *Invincible*, *Inflexible* and *Indomitable*. It will be remembered that they had recently been detached from Beatty in exchange for 5BS to undertake gunnery training. It was one of the supporting light

Fig. 5.10. The famous picture of Boy Seaman Cornwell, the
communications number on HMS *Chester*'s 'A' mounting.
It is somewhat idealized: the mounting had been hit and
Cornwell was to die of his wounds a few days later.

cruisers, the very new HMS *Chester*, that first made contact with the
enemy. She ran into a German light-cruiser squadron and was heavily hit.
A member of one of her guns' crews, Boy Seaman Jack Travers Cornwell,
won a posthumous Victoria Cross during this action for remaining at his
post. At 16 years of age, he is the third youngest winner of the medal.

Hood saw the gun flashes and turned to engage the four German
light cruisers, hitting three and crippling the SMS *Wiesbaden*. Hipper now
sent another light cruiser, SMS *Regensburg*, and no less than 31 destroyers
to attack the battlecruisers with torpedoes. They were met by the HMS
Canterbury, a sister ship of the *Chester*, and four destroyers. The German
ships fired 12 torpedoes and turned back, leaving HMS *Shark* damaged
and dead in the water without working engines. Her commanding officer,
Commander Loftus Jones, was fatally wounded after he replaced a casualty
serving one of the guns, winning the third posthumous Victoria Cross of
the battle.

Jellicoe was starved of information at this point. The commander of
his scouting force, Admiral Beatty, after his initial sighting report, had told
him nothing during the run to the south, or indeed during the run to the
north. Beatty did not communicate with his commander in chief from 16:45
to 18:06. While he had lost his long-range wireless, he could easily have
instructed one of the other ships to make a report, but did not do so. Indeed
Jellicoe did not even find out about the loss of two battlecruisers until the
next day. While Goodenough was sending him accurate reports, the position
that he was reporting did not tally with other reports. In the days before
accurate radio-based navigation aids, Goodenough would have given his
position derived from 'dead reckoning' (properly 'ded' from 'deduced'). At
its simplest, dead reckoning may be thought of as: 'I started at X, steamed
east at ten knots for an hour, so my position is now ten nautical miles east
of X.' Factor in the effects of currents, tides, repeated changes of course
and speed, and Jellicoe was faced with differences in reported positions of
as much as eight miles. Therefore, while he knew he was rapidly coming
on the *Hochseeflotte*, he did not know from which direction they would
appear. Jellicoe was faced with how to control an enormous battlefleet
moving at high speed and armed with long-range guns. His problem was
that he was steaming with his battleships in a cruising formation made up
of six parallel columns, each of four ships. If the Germans appeared ahead

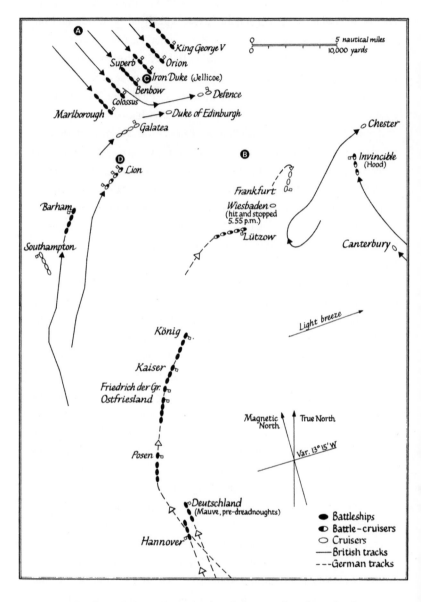

Fig. 5.11. HMS *Marlborough* sights HMS *Lion*. Admiral Hood and
3rd Battlecruiser Squadron engage the German battlecruisers.

of them in this formation, only the ship at the head of each column would be able to fire on them. If he changed into a line of battle on his present course by merging the columns, it would make the problem worse if the enemy appeared ahead; he would only have one ship with 'clear arcs' for its guns, and just for its forward turrets at that. If he deployed his columns into a line at an angle to his present course and went the wrong way, he might place himself steaming away from the enemy.

There had been a lot of thought given to this sort of situation before the war. Essentially, it came down to whether it was best to fight in a single line or in groups, termed 'divisions'. A further debate considered whether lines should be one ship after another, termed 'line ahead', or at an angle with one ship behind another, but with her course displaced to one side, described as 'deployment on a line of bearing'. The former meant that ships had to steam parallel, or nearly so, to their opponent; the latter allowed for converging or diverging courses, while still allowing the maximum numbers of guns to engage. The complicating factor in the North Sea was the visibility, which was frequently poor. Before radar, ships were restricted to how far they could actually see either from the bridge or up the mast, termed the 'fighting top'.[11]

Jellicoe opted for fighting in line ahead. That way, if a ship was not ahead of you or behind you she was enemy. Of course, errors could arise if the line became distorted, as was to happen to the US Navy off Guadalcanal in 1943.

At 18:00 HMS *Marlborough*, the ship at the head of the right-hand line, reported that *Lion* was in sight to the south-south-west, not directly ahead to the south-east as Jellicoe had anticipated. Jellicoe therefore altered the course of the fleet to the south. As soon as *Lion* could be seen from the bridge of the Jellicoe's flagship *Iron Duke*, Jellicoe asked Beatty by flashing light: 'Where is enemy's battlefleet?' Beatty actually did not know the answer, because he was not in touch with the enemy; 5BS astern of him was doing the fighting. So he didn't reply for five minutes, during which time Jellicoe's ships continued south as the two opposing fleets closed each other by nearly four miles. Beatty's eventual answer was unhelpful: 'Enemy's battlecruisers bearing south-east.' Jellicoe altered course again to the south-east. Jellicoe repeated his original signal and after seven further minutes received the reply: 'Enemy battlefleet in sight bearing south. The nearest ship is seven miles.' This was not a completely helpful response, but it at least gave

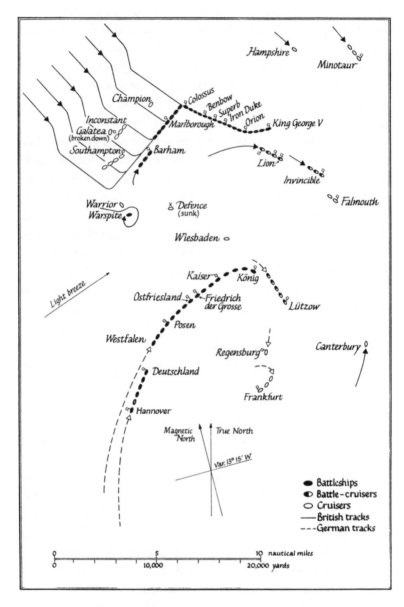

Fig. 5.12. 'Equal Speed, Charlie, London'. Jellicoe starts his fleet deployment to port. HMS *Warspite* gyrates and Admiral Arbuthnot commits suicide.

Jellicoe something to work on. He now had to decide how to deploy his ships. If he deployed the correct way as Scheer came into range, the Grand Fleet would be in a line at right angles across Scheer and his unsuspecting fleet; Jellicoe would have 'crossed his T', the ultimate aim of every admiral. His ships would be able to fire on Scheer's ships from either side with full broadsides, while Scheer would not be able to use all his firepower. If Jellicoe got it wrong, he would suffer the fate he wished for Scheer. His alternatives were many, but in reality he could either deploy to starboard or to port. He thought for less than a minute and told his fleet signal officer to deploy to port from the present course. His signal officer made a suggested amendment by deploying one point ($11.5°$) further to port to ensure that there was no possibility of misunderstanding. At 18:15, just in time, the signal ordering the formation, 'Equal Speed, Charlie, London', now famous among naval communicators, was made. Beatty and his remaining battlecruisers were to take station at the head of the line, and eventually Evan-Thomas and 5BS took station to starboard of the rear of the line.

As the Grand Fleet deployed into line, the battle orders laid down how the escorting cruisers and destroyers were to take up new stations while under German fire. This required consummate ship-handling; ships literally had to dodge in and out of the deploying battleships.

> [T]he scrimmage of all of us light craft ahead of the two fleets was a sight worth seeing [...] It will never cease to be a source of wonder to me that so few ships were hit and that there were no collisions. I think it must have been one of the most wonderful displays of seamanship and clear-headedness that ever existed.[12]

When Scheer emerged from the gathering gloom, the Grand Fleet was deployed across his path; he had had his 'T' crossed. As the German official history put it: 'Suddenly the German van was faced by the belching guns of an interminable line of heavy ships, extending from north-west to north-east, while salvo followed salvo almost without intermission.'[13]

There was an exception to the 'clear-headedness' displayed by most of the Grand Fleet. Stupidity combined with unnecessary courage was now to play its part. The 1st Cruiser Squadron (1CS) under Rear Admiral Arbuthnot had been in his screening position five miles ahead of the

Iron Duke. Arbuthnot caught sight of the German 2nd Scouting Group and turned *Defence* (his flagship) with *Warrior* to engage them. So single-minded was he that he cut across the bow of *Lion*, forcing her to turn to avoid a collision. Arbuthnot then happened on the crippled *Wiesbaden* and slowed to sink her, ending up less than 6,000 yards from the German battle line. *Derfflinger* and four battleships engaged Arbuthnot's ships, with the inevitable result. *Defence* was lost with all hands. Arbuthnot has been praised for his offensive spirit; he has also been condemned for his 'berserk rush' at the enemy, which not only cost him his ship and his life, but forced the *Lion*, then obviously in action with heavy ships, off her course, threw out the fire of her squadron, and made them lose sight of their target in the cruisers' smoke. An unkind critic might liken Arbuthnot's action to that of an undisciplined puppy chasing a kitten.[14]

Warrior was saved, albeit temporarily, by *Warspite.* As 5BS were getting into their proper station under enemy fire *Warspite*'s rudder jammed, and in front of the closing German fleet she turned two full circles, sustaining no less than 13 shell hits before getting her rudder back under control while also distracting the German fire from *Warrior.* When *Warspite*'s rudder again jammed she was ordered home; *Warrior* was also sent home, but sank while being towed.

Meanwhile, as the Grand Fleet deployed, Hood's 3BS had come upon the German battlecruisers and turned onto a parallel course at 9,000 yards. *Invincible* was shooting extremely well, hitting both *Lützow* and *Derfflinger* several times, but was being fired on in her turn by four battlecruisers and a battleship. A hit on 'Q' turret set off the charges within and *Invincible* blew up. One thousand and twenty-six of her ship's company died, leaving only six survivors. In fact there were more than that initially, but the battlefleet deployed through the survivors in the water and many were swept under by their bow waves and wash.[15]

Scheer was now in an extremely difficult position. He was heading directly for the British line, and thus the range was steadily and quite rapidly shortening. He could only engage the enemy over very limited arcs, while the British line was broadside on and, except for a few ships that did not have all-centre-line turrets, could bring literally every gun to bear on the German ships as they came into range. Scheer was about 150 miles from his bases, and he was hampered by Admiral Mauve's slower

Fig. 5.13. HMS *Invincible* sinking after a magazine exploded, taken
from HMS *Barham*. Ironically *Barham* was to be lost in World War II
after a magazine explosion, on that occasion due to a torpedo hit.

pre-Dreadnoughts. To make matters even worse, Jellicoe was heading to the
east as the ships deployed, meaning that his ships would be well placed to
position themselves between Scheer and Germany. If Scheer turned onto
a parallel course, he would be outnumbered and annihilated. Already the
British gunnery was beginning to tell; Hipper's battlecruisers at the head of
the line were taking the worst of it. *Lützow*, his flagship, was disabled with
no wireless; *Derfflinger* had a large hole in her bows; *Seydlitz* was flooded
forward and was heavily down by the bows; *Von Der Tann* did not have a gun
in action. To be able to command his ships Hipper had to move, that is 'shift
his flag'. He transferred to SMS *G39*, a destroyer, and in her went from
ship to ship before coming on the *Moltke*, which was largely undamaged.
The developing situation meant that he could not actually board her until
22:00. Because he was unable to exercise command from a destroyer, he
handed command of his ships temporarily to the captain of the *Derfflinger*.
The battlecruisers were not the only ships to take punishment: the four
ships of the leading battleship squadron had all been significantly damaged,
particularly *König* and *Markgraf*, the latter with her port engine out of action.

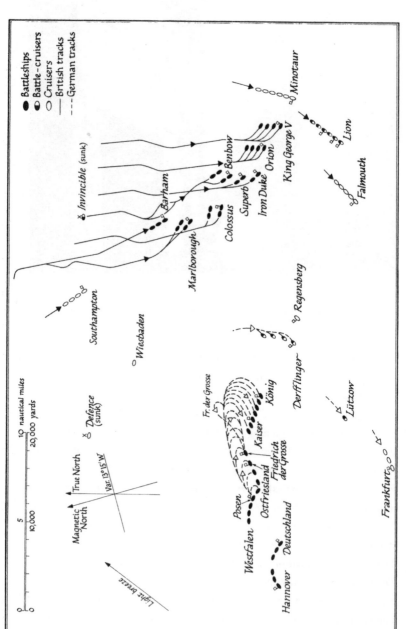

Fig. 5.14. *Gefechtskehrtwendung.* Scheer, with his 'T' crossed, turns away.

The reason that *Moltke* could not slow or stop long enough to let Hipper board was because Scheer had ordered *Gefechtskehrtwendung*: battle turn-away (the equivalent of the British 'Blue Pennant, One, Six'). Each ship was to turn to starboard in a half-circle, reversing the line. He ordered this at 18:33. Since the capture of the *Magdeburg* code-book, the British were well aware of this manoeuvre. It has already been asked why Beatty had not employed it at the end of the 'run to the south'; after all, such a manoeuvre was also in the British code-book. However, it was technically easier for the *Hochseeflotte* to do this. The British ships were at a spacing of two cables, 400 yards, which is actually very close together for fast-moving battleships that were themselves over a cable long. The German ships were spaced at three and a half cables, which made concentrating fire more difficult, but manoeuvring easier and safer.

Scheer covered the turn with a torpedo attack and smoke laid by the 3rd Destroyer Flotilla, steering south-west by 18:45. Almost immediately the *Hochseeflotte* was lost to the sight of the Grand Fleet. Jellicoe did not anticipate that the Germans intended to break to the west into the open sea. He chose instead to turn so that the Grand Fleet would block Scheer's route to Wilhelmshaven. He first altered course to the south-east and then south. At this point one of his battleships, *Marlborough*, was hit by a torpedo but remained in her station. After the changes of course, the Grand Fleet was now in six columns, steering parallel courses, but with the divisions not overlapping.

One of the disputes about Jutland is the action of Admiral Beatty at this point. Should he have resumed his scouting role or, the Grand Fleet having deployed, remained as the leading element of the fleet? He chose the latter, but for a reason never explained led his ships in a full circle, taking about 15 minutes. This had the effect of putting his ships further away from the Germans than the remainder of the Grand Fleet. Fortunately Goodenough was 'the one light cruiser squadron commander who appreciated that his reconnaissance duties were not necessarily over [when the fleet deployed] but that whenever contact was lost, he must push in and obtain information for the C in C'. Goodenough continued reporting and informed Jellicoe of Scheer's next turn.[16]

The next development has also caused much debate, but this time on the German side. Scheer carried out another 16-point (180°) turn. His account suggests that he felt he would be able to strike suddenly at the British fleet; the German official history actually draws a parallel with Nelson's tactics at Trafalgar, but by very selective quotations. In any event, Scheer first encountered the two rear divisions of the Grand Fleet, but then rapidly came under fire from the entire battle line. During this 'second contact' the *Colossus* was hit twice without sustaining significant damage. She thus had the distinction of being the only Grand Fleet, as opposed to BCF, battleship to be hit by German shellfire at Jutland. German ships were hit 27 times, of which 19 hit the battlecruisers. Realizing that he was in an impossible situation, Scheer ordered his battlecruisers forward to cover an intended withdrawal. Some authors have called this their 'death ride'. They had closed to within 8,000 yards of the British line when Scheer called them back and then, covered by a destroyer attack combined with a smokescreen, ordered yet another 16-point turn away from the Grand Fleet.

Jellicoe's response to turn away from the torpedo threat is one of his most contentious actions of the day. His critics included Beatty and the naval historians (themselves naval officers) Roskill, Dewar and Richmond, as well as Winston Churchill. Many who were present apparently expressed 'surprise' retrospectively. They held that Jellicoe should have turned towards the enemy. Churchill went further, arguing from a map drawn after the war showing the positions of the two opposing fleets (which Jellicoe did not know at the time), that Jellicoe should at this point have split his fleet into two 'and so take the enemy between two fires'.

It has already been mentioned that the Grand Fleet ships were steaming close to each other and that the gaps between them were relatively small. Jellicoe was anticipating a mass attack, possibly in waves from different directions, which is what Scheer should have ordered. Jellicoe had to turn either towards or away, because if a destroyer attack came at right angles to the British line, the fleet would have presented a target that would have been roughly equal ships and gaps. Even fired randomly, torpedoes would have had a 50 per cent chance of a hit. With his ships steaming away from the launching ships at 20 knots, and with German G6 and G7 torpedoes being capable of 35 knots at short range and about 28 knots at longer range, he would have had a reasonable chance of outdistancing them when their fuel ran out, or of dodging them because of the much slower relative speed. The alternative was to turn towards, making for a closing speed

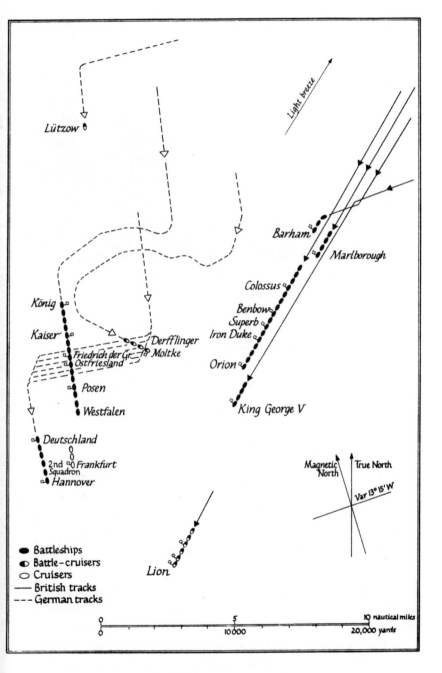

Fig. 5.15. Scheer turns away again.

Fig. 5.16. German destroyers attacking. The Germans know the
battle of Jutland as the *Skagerrakschlacht*, or 'Skagerrak Battle'.

of about 50 knots, which would have been suicidal given the difficulty
of seeing an approaching torpedo. His decision to turn away should not
have come as a 'surprise' to his subordinates; it was the prescribed tactic
to be employed in the circumstances in the Grand Fleet Battle Orders,
and was the standard tactic in the Royal Navy as well as in the German,
French, Italian and American navies. In the event, *Marlborough*, already hit
earlier by one torpedo, avoided three further; *Revenge* 'swerved' to avoid
two; *Hercules* and *Agincourt* altered course to avoid torpedoes; one passed
between *Iron Duke* and *Thunderer*; *Colossus* avoided one, as did *Collingwood*.
Four more were seen from *Barham*. *Neptune* saw a torpedo coming from
directly astern, too close to avoid it, but nothing happened; presumably
the torpedo ran out of fuel at the last second. The turn away also prevented
some German destroyers even from launching their weapons, and the
attack cost the Germans five destroyers.[17]

Chatfield, who was Beatty's flag captain and was later to be the best
First Sea Lord of the interwar years by far, said: 'Most experienced
Commanders would probably have acted as did Sir John Jellicoe.' In par-
ticular, the criticism of his action came particularly oddly from Beatty,
who had turned away from the mere sight of a periscope at the battle of
Dogger Bank.[18]

Anticipating that he would come on the enemy, Jellicoe now steered south-east and gradually worked his course round to south. Goodenough reported that the *Hochseeflotte* was steering west. Jellicoe was now between Scheer and Germany, and worked his course incrementally more westward as darkness came closer. Beatty now intervened. He was to the east of Jellicoe, but signalled to Jellicoe: 'Submit van of battleships follow battle-cruisers. We can then cut off whole of enemy's battlefleet.'

This signal, like much of what happened that day, has been seized on by both Beatty and Jellicoe protagonists. In naval usage, 'submit' to one's superior is insubordinate; indeed Jellicoe later recorded that he thought it was. The Beatty camp felt that the signal was the thrusting subordinate trying to urge on an indecisive superior; Jellicoeites thought that it was posturing, since at the time the signal was made Beatty wasn't even in sight of the enemy. In fact, parts of the Grand Fleet were actually closer to the enemy than Beatty, and by the time Jellicoe received the signal he was already steering closer to the enemy than Beatty himself. However, Jellicoe assumed that Beatty must be in sight of the Germans, and ordered Admiral Jerram commanding the 2nd Battle Squadron (2BS) to follow Beatty at 20:07. At 20:12 Beatty came on the German battlecruisers. *Seydlitz* was hit again, and the intervention yet again delayed Hipper, who was at this moment still trying to board the *Moltke*. The 'five-minute squadron' of German pre-Dreadnoughts intervened, and three in their turn were hit, but as they turned away, having distracted Beatty from the battlecruisers, Beatty did not follow them.

The night of 31 May–1 June

Jellicoe was not keen to fight a night action. He probably overestimated his opponent's capability to do so, ascribing to the Germans the ability to use searchlights to find targets, coordinated with gun's crews trained in night firing. His major concern was that the Germans would launch night-time destroyer attacks with torpedoes and that the secondary armament of his ships, that specifically designed to counter destroyer attacks, was not yet director-controlled. In essence, he was concerned that his advantage of numbers and firepower would be negated: 'It would be far too fluky an affair.' The reality was that all Scheer wanted was to get home: his battlecruiser force was spent, he had twice had his 'T' crossed, and those of his Dreadnoughts that had come under British fire had been battered. The Germans were in no mood to fight a night action, and when during the night one was effectively offered to them on advantageous terms, they declined it. Jellicoe did not fully realize it, but the German fleet was beaten and trying to get away. He knew, however, that he was in a position to block their escape, but was in a quandary as to which route home they were likely to take. Viewed from *Iron Duke*, Jellicoe's flagship, there were four routes open to Scheer. He could conceivably have gone to the north, around the tip of Denmark, and escaped into the Baltic. This was impractical: it was far too far to take the damaged ships Jellicoe knew the Germans had. The second possibility was to steer south-east towards the Horns Reef light vessel, then into the Jade River. Jellicoe discounted this as he felt that Scheer had attempted it twice during the day, and he did not believe that Scheer would try it again by night, thinking it most likely that he would attempt one of two possible southerly routes, despite these being constrained by British minefields.[19]

While Jellicoe thought a night action unlikely, he remained concerned that the Germans might attempt destroyer attacks; for the four hours of darkness, he ordered his ships into three columns, each one mile apart, and placed his fifty-eight destroyers five miles astern to protect the battlefleet from just such an attack. He then steered south. Scheer at this time was about ten miles north-west. He decided that he would try to slip behind the Grand Fleet, accepting if necessary that he might have to fight his way through to the Horns Reef.

During the short summer hours of darkness there was a series of actions between smaller ships. First Goodenough in *Southampton* with *Dublin*, *Nottingham* and *Birmingham* in company found themselves steering parallel at 800 yards to a line of unidentified cruisers. *Dublin* fired on them and the German cruisers fired back. *Southampton* lost 45, killed, and her all-important wireless. However, she fired a torpedo that sank the *Frauenlob* with all hands. The Germans broke off the action.

There followed a series of at times desperate actions between British destroyers and cruisers and German battlecruisers and battleships. *Black Prince*, originally part of Arbuthnot's 1st Cruiser Squadron, had become detached from her squadron during the deployment and missed the fate

Fig. 5.17. British destroyers, in this case part of *Barham*'s screen.

of *Defence* and *Warrior*. However, having initially turned north-west, she then steamed south in search of the British fleet. At about midnight, she encountered a line of Dreadnoughts and challenged them. Unfortunately they were German and opened fire at point-blank range. *Black Prince* was lost with all hands.

Of the destroyers, firstly *Tipperary* saw what she initially thought was a British light cruiser but, when she challenged it, was instantly fired on by a light cruiser and two Dreadnoughts, the *Rostock*, *Westfalen* and *Rheinland*. She engaged with torpedoes, but a German salvo killed or disabled every person and gun in the fore part of the ship. Acting Sub-Lieutenant N. J. W. William-Powlett continued to fight the ship from aft, but she was soon to sink. *Spitfire*, next astern of her, engaged what her captain took to be a cruiser. It was actually the battleship *Nassau*, and *Spitfire* was rammed for her temerity. With her captain now incapacitated, the first lieutenant (second in command) took over and managed to avoid being rammed again, this time by either *Derfflinger* or *Seydlitz*. When *Spitfire* eventually made it back to the Tyne, the shipyard found 20 feet of the *Nassau*'s plating and part of her anchor gear wedged in her forward mess deck![20]

The continuous series of destroyer attacks caused some confusion among the German ships, which led to the battleship *Posen* ramming and sinking

the light cruiser *Elbing*. At about the same time, *Rostock* was torpedoed and
was later to sink, but which ship fired the torpedo is uncertain. British ships
were not immune from self-inflicted damage: *Sparrowhawk* was rammed
forward by the *Broke*, losing her bows, and aft by the *Contest*, losing part
of her stern. She was to sink, but not before picking up *Tipperary*'s survivors and having the surreal experience when at around dawn she sighted
a German cruiser. The crew of *Sparrowhawk* stood by their remaining guns
as the cruiser came closer; she was noted to be listing and sinking by the
bows, and then she 'quietly stood on her head, and sank'. This was probably the *Rostock*, which had already been abandoned by her crew after
being torpedoed.

Sir Julian Corbett, the official historian writing of this phase of the
fighting, praised the 4th Destroyer Flotilla and concluded with an important point:

> Alone they had borne the brunt of the whole German battle fleet, and not
> a man had flinched. Again and again as a group of the enemy tore them with
> shell at point-blank and disappeared they sought another, and attacked till
> nearly every boat had spent all her torpedoes or was a wreck. Such high
> spirit and skill had they shown that one thing was certain – the failure of
> the flotilla to achieve all that was generally expected from it was due to no
> shortcoming in the human factor. It was the power of the [torpedo] itself that
> had been overrated.[21]

What was extraordinary was the inaction of the main body of the Grand
Fleet. Numerous ships saw and heard the series of actions, yet did nothing.
Not only did they not investigate, but they did not report to Jellicoe, who
was left in blissful ignorance of the *Hochseeflotte* passing astern.

Probably the worst example of what can realistically be described
as neglect of duty was the saga of the *Seydlitz*. She had been ordered to
make her way independently back to Germany. Without a gyrocompass
and with damaged charts, she passed down the starboard side of the
British battle line. Three ships, *Agincourt*, *Marlborough* and *Malaya* all saw
her from as close as 4,000 yards and did nothing; they did not engage,
challenge or report her, and *Seydlitz* eventually made it home. It would
appear that admirals and captains thought that 'a stream of wireless reports

Fig. 5.18. SMS *Seydlitz* returning to Germany after the battle, barely afloat.

in company [...] seemed superfluous and uncalled-for'. Throughout the night it would appear that everybody thought somebody else would report, and nobody did. There was a notable exception, Captain Stirling of HMS *Faulknor*. He reported the position, course and speed of the German fleet three times, but the signals appear not to have got through. Then he and his six destroyers attacked, firing 17 torpedoes. They missed the Dreadnoughts, but sank the SMS *Pommern*. That and the mining of the SMS *Ostfriesland* on a mine laid earlier by HMS *Abdiel* were the last actions of the battle.[22]

During the night actions, the British destroyers had sunk a pre-Dreadnought battleship, three light cruisers and two destroyers at the cost of five destroyers. *Lützow* had been left to her fate and sank in the early hours, completing the German losses. It is of interest, if only to indicate how little the submarines of both sides played in the battle, that Scheer and the *Hochseeflotte* had a marked attack of 'periscopitis' as they neared home.

In the morning, informed by the Admiralty that the German fleet had returned home, and realizing that he was close to the German coast and had been seen by a Zeppelin, Jellicoe returned his ships to their bases.

Who won?

More has been written about the battle of Jutland than any other sea battle
in the twentieth century. It has proved to be an endless source of controversy
and debate, not only centring on 'Who won?', but also involving subsidiary debates: 'Who performed well or badly?', 'What went wrong?' and
'Whose fault was it?' Unfortunately very early on the debate came down
to personalities, characterized as the Jellicoe–Beatty argument, which
came close to splitting the navy in the 1920s. Beatty, who was to be First
Sea Lord from 1919 until 1927, was well placed to influence the debate,
and even to force through his interpretation of events, while Jellicoe's
reputation suffered at his hands, and indeed those of Lloyd George and
Winston Churchill.

After the battle, the *Hochseeflotte* returned to Germany and loudly
proclaimed their victory, based on numbers of ships sunk. The Admiralty
by contrast delayed any release of information until the Grand Fleet was
back in Scapa Flow, which inevitably took longer, and then produced a
factual statement without interpretation. The stage was set for the Jutland
controversy. To this day some German sources claim the *Skagerrakschlacht*,
or 'Skagerrak Battle', as a German victory. The Admiralty were slow to
realize the need to manage the media better. They were always, in the
modern idiom, playing catch-up, and they were not helped by British
introspection and the blame game.

The criticisms of the British performance fall into two groups: insti-
tutional rigidity and incompetence by commanders, and material failings.
Since many authors have used the battle of Jutland as being indicative of
the performance of the Royal Navy in the totality of World War I, it is
appropriate to address the matter here.

Looking at the battle of Jutland, many have felt that it epitomizes a
rigid, unthinking officer corps up to and including admirals, in thrall to
the concept of 'all laws are as naught beside this one, / thou shalt not
question but obey'.[23]

The typical naval officer, if such a man existed, has been described as 'an
automaton who only came to life at the impulse of a superior'. Churchill
was another to make much of the supposed lack of initiative shown by
naval officers: he said, 'Everything was centralized in the Flagship, and all

initiative except in avoiding torpedo attack was denied to the leaders of squadrons and divisions.' Churchill went further, attributing their poor performance to poor education, 'mere sea service'.[24]

What most writers tend to forget or ignore is that naval officers have not only to obey orders instantly and even unthinkingly, but if necessary to display immediate initiative. Instant obedience is essential because 'things happen too quickly at sea to allow time for long and detailed instructions. Orders must be short and snappy, and they must be instantly and exactly obeyed.' Discipline at sea rests on that very basis, and often there can be no room for individuality or initiative. The management of ammunition in a gun turret has already been discussed. Looking in a bit more detail at the British twin 15-inch BL Mk I magazine/shell room/barbette/turret in the 5th Battle Squadron ships of the *Queen Elizabeth*-class, it was very similar to the guns of the other British Dreadnoughts, in that each was manned by 64 men. Each man had particular duties and actions to be carried out in a particular and specific sequence: 37 actions to load and fire, seven further after firing. If these were not carried out in the correct sequence, each gun could not and did not fire continuously. This required discipline, obedience and teamwork of the highest order. This was no place for initiative; indeed, as has been shown, cutting corners and circumventing regulations may well have caused the loss of three battlecruisers. It may be wondered why, when it was known that regulations were circumvented, the navy was prepared to allow the myth to grow up of poorly designed battlecruisers. It is probable that a tacit decision was taken that blaming the material was better than blaming the men.[25]

Separately, as soon as there was action damage, death of a senior officer or any other change, initiative in circumstances was required, and at Jutland this was displayed in abundance, particularly in the destroyers. From among these officers came some of the great naval leaders of World War II, including Admirals Tovey, Cunningham and Vian. Nonetheless, there were many examples of lack of initiative, notably by officers who were the product of the period of rapid expansion in the navy.

It was said at the beginning of this chapter that discussion of the battle of Jutland became very polarized. For nearly a century there has not been a generally accepted view of what went right and what went wrong at the battle without it being taken as 'pro-Beatty' or 'pro-Jellicoe'. Certainly rational discussion was precluded by the machinations of Beatty as First Sea Lord from 1919 to 1927, who took charge of the production of the official account of the battle. A 'pro-Jellicoe' book, *The Truth about Jutland* by Rear Admiral Harper, appeared in 1927, and since then there have been numerous accounts. Some have looked at material 'failings'. The supposed design flaws, indeed the whole concept, of battlecruisers has been a fertile source of argument, most recently in 2012. The failings of British armour-piercing shells have been laid at Jellicoe's door because he had been the director of naval ordnance. Indicative of the partisan nature of the arguments, Beatty and Chatfield ignored the fact that Jellicoe had requested trials of the shells against armour before leaving the post; his successor had not followed it up.[26]

Nearly a century after the battle it should be possible to make a more balanced judgement. What is certain is that nobody had fought a battle like it before or since on the same scale. The 1905 battle of Tsushima between the Japanese and Russian navies actually gave little indication of what was to come, and there has only been one subsequent battle in which Dreadnoughts formed a line of battle, the battle of the Surigao Strait in 1945.[27]

Presciently, Jellicoe discouraged the use of wireless. While wireless technology steadily improved in terms of both range and speed, he was concerned with security, in terms of betraying both position and content. However, while his Grand Fleet Battle Orders firmly stated that it could be used in sight of the enemy, many avoided doing so; indeed the escape of the *Seydlitz* was probably due to the avoidance of using wireless, as well as an astounding lack of initiative.

Those who encountered the *Seydlitz* were far from the only ones; many others failed in their duties to report. Everybody knew that the commander in chief maintained a plot and had the best picture of what was happening, but they forgot that it was the information they as subordinates provided which enabled the picture on that plot to be compiled. The only subordinate commanders to perform their duties to report consistently, and who reported useful information, were Commodore Goodenough of the 2nd Light Cruiser Squadron and Captain Stirling of the 12th Destroyer Flotilla. Commodore Goodenough not only had what would now be called a good command team, he delegated within it and exercised it in harbour. Beatty manifestly did not have a professional command team. To spend the amount of time he did on composing his initial sighting report and delaying opening fire suggests a concern for detail management. Seymour let him down, but that was Beatty's fault; Beatty had chosen him and kept him after earlier failings.[28]

As for performance, the main body of the Grand Fleet made very good shooting and in turn returned home almost undamaged. They shot well and accurately, and inflicted a lot of damage. The Battlecruiser Fleet did not, however, and knew that they would not even before the battle. In an effort to compensate for their inadequate shooting, they broke the rules of magazine safety. Initiative is only laudable if it works; in this case it cost the navy three ships. There is no need to invoke any other cause for the loss of the battlecruisers, although it could be argued that Beatty thought that he and his ships were too big and important for a scouting role and that they belonged in the battle line. They did not: they were scouting ships.

Jutland showed that the torpedo had been overrated as a weapon. Before the war, it was seen as the battle winner; indeed there were some, such as the French *jeune école*, who felt that it was all a navy needed, and some have argued that Fisher was moving towards this way of thinking. It proved to be a useful adjunct, but it was not the decisive factor that had been feared to the extent that it drove tactical thinking, and it was not a battle winner.

Were the people of the Royal Navy inadequate? Were they automata? Had they received too technical an education, or had they been educated by 'mere sea service' with an inadequate broader knowledge? The reality was that the generic naval officer had breadth and was, as has been said, rigidly disciplined but showed great initiative when necessary. However, the Royal Navy was to an extent let down by the quality of some, who in an average year, let alone a bad one, would not have made selective promotion. Arbuthnot comes to mind: brave to a fault, but not very bright.[29]

So who did win? An oft-quoted American journalist wrote after the battle that 'the German navy has assaulted its jailer, but remains in jail', which is a fair summation. Battles aren't won by accountants totting up numbers of ships and men lost. On land the winner of a battle is the army left in possession of the battlefield. In reality, the battlefield of Jutland was the North Sea. All of the admirals who commanded the so-called 'High Seas Fleet' spent their time trying to avoid fighting the Grand Fleet, which meant staying in harbour. Scheer was no exception; having encountered the Grand Fleet at Jutland, he spent almost his entire time trying to get away. The Royal Navy never came close to losing control of the North Sea, although the German surface fleet could still play in their prison exercise yard in the Heligoland Bight and the Baltic. By far the furthest the *Hochseeflotte* ever ventured in wartime was to Scapa Flow to surrender.

CHAPTER 6

The Navy and New Technologies

The New Navy took its historical place
In warfare on quite unconventional lines…

Rear Admiral Ronald A. Hopwood[1]

It is difficult 100 years later to comprehend fully how many technologies that are now considered normal parts of our civilization had come into being or to fruition in the years immediately before and during World War I. The internal combustion engine was already well known, if not widely used, but wireless, the heavier-than-air aeroplane and, in the naval area, the submarine and the torpedo were still in the very earliest phases of their development.

Before and during World War I, the Royal Navy not only led the development of technologies such as wireless, but rapidly evolved methods of using them (what would now be termed 'doctrine'). It was the source of many innovations that were to change the face of warfare, not only at sea, but also on land and in the air. It was by far the most fertile source of innovation among the Allies, and indeed among the Central Powers. It was a vibrant and confident service that encouraged new thinking and was far from being the hidebound service resistant to change described by many.[2]

The armoured car and the tank

The Royal Naval Air Service in the first days of the war was deployed into Belgium, where it used a variety of touring cars, including Rolls-Royces, Lanchesters and, ironically, Mercedes. These were used to pick up airmen who had been forced to land, which included many in territory already overrun by the Germans. Inevitably, while carrying out such rescues, they sometimes came under fire. Their commanding officer, Commander Charles Samson, one of the pioneers of naval aviation, armed one of these vehicles with a Maxim gun and actually ambushed a German car on 4 September. Having been armouring its ships for much of the preceding century, armour was part of the naval way of thinking in a way that it was not for the army. The army did not begin to use any form of armour, and only steel helmets at that (apart from simple shields on field guns), until the war was a year old; until then soldiers fought in soft caps. Samson had the local shipbuilders apply boilerplate to his vehicles, and the armoured car was born. Samson requested that the Admiralty should take this further, and they asked Rolls-Royce to build further cars with armour plate and a gun in a revolving turret. The concept was taken further, using the chassis of an American lorry and mounting a 3pdr gun, known as the Seabrook.[3]

Within months there were twenty squadrons of armoured cars, formed into the Royal Naval Armoured Car Division (RNACD). These were used as mechanized raiding columns against the German forces during the retreat from Antwerp to the River Yser. Some of the Seabrooks fought at the battle of Neuve Chapelle. RNAS armoured cars were even attached to the 3rd Cavalry Division in IV Corps under General Rawlinson. As the war became less mobile with the development of trench warfare, there was little use for armoured cars on the Western Front (or indeed for the cavalry they were supplementing), and they were redeployed, some to Gallipoli, some to the Middle East, to be used in the desert war against the Ottoman Empire. The army took over control of most of the naval armoured cars, and in the summer of 1915 the RNACD was disbanded.

An army officer, Lieutenant Colonel Swinton, was an official cor-respondent with the army and had seen the armoured cars at work. He wondered whether a marriage of armour and the tracked tractor used in agriculture might be the answer to the increasing stalemate in the trenches. As a uniformed war correspondent he had significant access, and so he put the idea to Maurice Hankey, the secretary of the Committee on Imperial Defence. The army were less than impressed with the idea. However, Churchill got to hear of it, and set up the Admiralty Landships Committee under Tennyson D'Eyncourt, the Director of Naval

Fig. 6.1. Rolls-Royce armoured cars of 8 Squadron Royal Naval Air Service.

Fig. 6.2. Seabrook armoured car of 1914. This was based on
an American lorry chassis and carried a 3pdr gun.

Construction. Meanwhile, Commodore Sueter, responsible for the RNAS in the Admiralty and for the armoured cars, had been thinking much along the same lines. When the RNACD was disbanded, he retained 20 Squadron and began experiments aimed at what amounted to a land battleship with 12-inch guns. In the face of army disinterest, the Admiralty was about to abandon its efforts when the losses in the battles of early 1915 led to the army actually taking interest in and then taking over the development of what was to become the tank. It is fair to say that without the initial impetus from the RNAS and the Admiralty, the story of the tank's inception would have been much different.[4]

Air power

The first powered flight had been made in December 1903 by the Wright brothers in the United States. Powered heavier-than-air flight was therefore less than 11 years old at the outbreak of World War I. Initially it was not seen to have any military use. Indeed, having been turned down by their own government, the Wright brothers offered to sell their patents to the British Admiralty, but were again rebuffed by the First Lord. Admiral Fisher, the First Sea Lord, ever the visionary, grasped the potential of air power and obtained funding to build an airship. When Churchill became First Lord he took up the baton and was a naval air power believer, even training as a pilot himself. The result was that by 1915 all of the elements of modern air power were already recognizable, and it is notable that most of those and many subsequent developments were due to or had come about under the aegis of the RNAS.[5]

On 25 April 1912 the Committee on Imperial Defence established the Royal Flying Corps (RFC), with a naval and military wing. The naval wing rapidly developed a separate existence as the RNAS. The airship ordered by Fisher was less than successful. Presciently named the *Mayfly*, it did not fly, breaking its back while still moored to the ground. While the navy was to return to airships later in World War I, it now concentrated on heavier-than-air flight for what was then termed 'aerial navigation'.

In 1910 an American, Eugene Ely, had taken off from the USS *Birmingham* while she was at anchor; in 1911 Lieutenant Samson, who was to be one of the leaders in the development of naval aviation (and armoured cars), took off in a Short *S.27* from the deck of HMS *Africa*; the following year an aircraft took off from HMS *Hibernia*, a battleship, while she was underway. By 1912 the Admiralty had already determined that it needed two types of aircraft: one for scouting and one for attack or strike. It carried out trials of dropping bombs from aircraft the same year and immediately realized the difficulties of aiming them. Before the outbreak of war, it set up a research-and-development section to develop a bombsight and to determine the best way to attack a submarine. By the outbreak of war it not only had a viable torpedo-dropping aircraft, it had also begun to look at using aircraft-based wireless. The navy was very much aware of the potential of air power and its breadth.

The Royal Naval Air Service goes to war

With the outbreak of war, the army's Royal Flying Corps moved almost in its entirety to France, leaving behind only a few training machines. On the Western Front the RFC concentrated on reconnaissance, only developing fighter aircraft, initially termed 'scouts', to protect them. That was at the time all the army saw itself as needing; it was the RNAS that was to develop other dimensions of air power.[6]

In September 1914 an RNAS squadron equipped with Sopwith *Tabloid* aircraft was deployed to Ostend to protect the British Expeditionary Force from Zeppelin raids. At the same time there was in the public mind a developing fear of German airships, Zeppelins. In fact the German army used Schütte-Lanz airships, but the term 'Zeppelin' was widely used to describe any German or Austrian airship.

There was a very great keenness among Royal Navy commanders to take the war to the Germans, hence the wish to attack the airships in their bases. This would reduce the scouting ability of the German navy, as naval airships were employed in this role as well as in bombing civilian targets. Because Zeppelins were particularly vulnerable to the effects of weather when on the ground, they were protected from the elements in large hangars. Their second vulnerability was that they used hydrogen gas as their lifting agent, a very inflammable gas. Thus attacking them on the ground where they would be a static target was seen as being a good way of dealing with the Zeppelin threat.

Sopwith *Tabloid*s could carry two 20-pound bombs, and on 22 September the RNAS launched a land-based attack on the Zeppelin shed at Düsseldorf. The raid was unsuccessful, but a follow-up raid in November successfully destroyed the shed and the Zeppelin *Z9* that was inside it. Another raid, this time by RNAS Avro 504 aircraft, successfully attacked the Zeppelin factory at Friedrichshafen.[7]

The Cuxhaven raid

What is now remembered as the first air strike from the sea was an attack on German airship sheds at Cuxhaven. The first attempt took place on 25 October 1914. The attack was carried out by the Harwich Force and supported by the 1st Battlecruiser Squadron (1BCS) in case German surface units tried to intervene. Bad weather prevented the aircraft from being launched, so a second attempt was made two months later on Christmas Day. The attack itself was made by nine seaplanes launched by the *Engadine*, *Riviera* and *Empress*; it was supported by the Harwich Force cruisers and destroyers and, on this occasion, the entire Grand Fleet waiting in the middle of the North Sea. The raid itself was unsuccessful as the aircraft could not find the airship sheds and unsuccessfully bombed what would be later called secondary targets, German warships. However, this was the first seaborne air attack in history and it was an omen for the future. It was to be followed by far more successful raids as the RNAS extended its reach from land and sea.[8]

Aircraft

The pressures on the RNAS led to the development of aircraft that were much different from those required by the RFC. For example, initially the army saw no need for an aircraft capable of carrying bombs, whereas the RNAS did. From the start, the RFC bought its aircraft from the Royal Aircraft Factory at Farnborough; this was run by the Civil Service and was extremely bureaucratic. The designation of aircraft types was decided by civil servants; even their own designers did not understand the system. This was not surprising, as each aircraft they designed and produced was grouped according to which French aeroplane it vaguely resembled; thus the 'Farman Experimental 1' was designated *FE1*. In May 1914 the *BE2c* (Blériot Experimental) from the Royal Aircraft Factory was chosen as the standard equipment for both the RNAS and the RFC. However, the Admiralty chose to have its aircraft built by a private contractor, and thus began a long relationship between Blackburn's and the Royal Navy. The Admiralty policy of going to industry rather than relying on government design and manufacture led to innovative design and procurement, unlike that undertaken for the RFC. Eventually the RFC was supplied with some good aircraft, but only when it did what the Admiralty had done much earlier: go to private enterprise.[9]

Another Admiralty contractor for the *BE2c*, William Beardmore & Co., sought permission to develop a better aircraft. Unfortunately the aircraft they developed (the *WBII*), while a good aeroplane, used the same Hispano-Suiza engine as the Royal Aircraft Factory *SE5a*, which was given priority. This was actually quite ironic, because the Royal Aircraft Factory had declined to order the Hispano-Suiza engine. The Admiralty had acquired a licence to manufacture it, which was taken over by the Royal Aircraft Factory at government direction. Beardmore was an extremely innovative firm, which went on to modify Sopwith *Pup* shipboard fighters to have folding wings, in order to make better use of ship-borne hangar space, and a retractable undercarriage, although the latter was not successful. They also built the *WBV* to meet an Admiralty specification for a fighter armed with a 37-mm cannon intended as a Zeppelin-destroyer; this was at a time when the normal aircraft armament was two .303-inch (7.7mm) Vickers or Lewis machine guns. Of even greater interest was another 'Zeppelin-killer' built to meet an Admiralty requirement. The firm had started as Pemberton-Billing, named for its founder. He was bought out in 1916 and the firm became Supermarine. Its aircraft built to meet the specification never reached service, 'but contained several remarkably advanced ideas'. One of the design team was R.J. Mitchell, who as Supermarine's head designer was to be responsible for the Spitfire. The Admiralty fed off these and other firms, and they in turn fed off the Admiralty. Without that impetus they would not have developed into the innovative firms they became.[10]

Fig. 6.3. Land-based Royal Navy anti-aircraft gun.

Home air defence

With the majority of the RFC in France, responsibility for home defence – that is, air defence of the United Kingdom itself – was passed to the RNAS. The Admiralty addressed this new role in imaginative ways, not seeing it as being totally reactive to an enemy threat, but rather looking for ways to defend the country by offensive measures. Ground-based defences were not neglected, and the navy created in October 1914 the Royal Naval Volunteer Reserve Anti-Aircraft Corps, which operated both guns and searchlights. Initially these were used for home defence, but within months foreign service sections were created and went overseas to France and the Dardanelles.[11]

The threat initially came from Zeppelin rigid airships, because the Germans did not have a strategic bomber until 1917. The Admiralty soon realized that standing patrols along the coast to intercept raids was unproductive. However, the Zeppelins flew at over 10,000 feet, and initially there were no aircraft that could climb fast enough to be able to react to an attack, so a lot of effort went into developing aircraft that could. Even before better aircraft became available, attempts were made to attack Zeppelins in the air. After unsuccessfully attacking a Zeppelin with a machine gun on 17 May 1915, Lieutenant Warneford Royal Navy bombed *LZ 37* in the air on on 7 June 1915, destroying it and winning a Victoria Cross.

Responsibility for the air defence of the United Kingdom was passed back to the RFC in 1917. This was soon after General Haig had requested as a matter of urgency 'a very early increase in the numbers and efficiency of the fighting aeroplanes' in France. He had actually asked for 20 more squadrons. The consequence was that the Admiralty transferred substantial numbers of aircraft and engines to the RFC and deployed five home defence squadrons of its brand new fighter the Sopwith *Triplane* to France. The result was that they were not in England when the German strategic bombing attacks with *Gotha* bombers started in 1917. As will be described later, the absence of aircraft for home defence was one of the factors that led to the formation of the Royal Air Force.

The Western Front

Confusingly, the RNAS squadrons sent to France were numbered with
no regard to the numbers allocated by the RFC to their squadrons. Thus
in France there was now to be a 1 Squadron RFC and 1 Royal Naval Air
Squadron. This, along with what was felt by the naval pilots to be RFC policy
to minimize their contribution, has meant that the performance and even
the presence of the RNAS in France remains largely unrecorded to this day.

Five squadrons of Sopwith *Triplanes*, known as 'Tripehounds' or 'Tripes',
were sent to France. This aircraft was only used by the RNAS. They were
remarkable in many ways, but most particularly for their rate of climb;
they could reach 15,000 feet in 19 minutes, which made them eminently
suitable for air defence. In France they proved to be formidable fighters; as
well as their high rate of climb, they were extremely manoeuvrable. During
a service career of only seven months they established an 'enviable combat
record'. This was a welcome lift to British morale after 'bloody April', when
the RFC had suffered appalling losses. The RNAS, unlike the RFC at this
time, allowed squadrons and pilots to 'personalize' their aircraft. 'B' flight
of 10 Squadron RNAS, led by Raymond Collishaw, had black cowlings on
the aircraft of the so-called 'Black Flight' named *Black Roger*, *Black Prince*,
Black Sheep and Collishaw's aircraft *Black Maria*.[12]

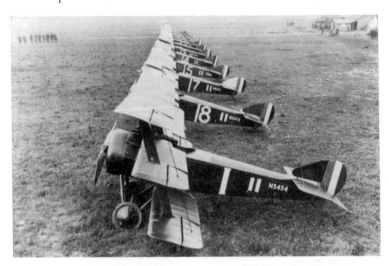

Fig. 6.4. Sopwith *Triplanes* of 1 Squadron RNAS in France.

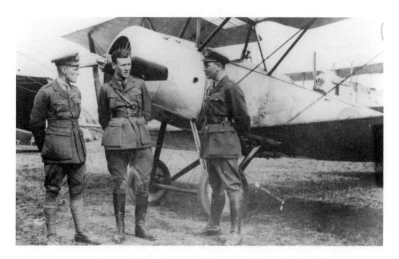

Fig. 6.5. Raymond Collishaw (right). His naval rank insignia can
just be made out on the cuff of his army-style uniform.

In 27 days the 'Black Flight' scored 50 victories, 16 to Collishaw,
without losing a plane. Collishaw went on to achieve 61 victories, one of
the highest-scoring British Empire aces. He later flew the famous Sopwith
Camel, which was first used by the RNAS, but he remains almost unknown
alongside the RFC aces such as Mannock, McCudden, Bishop and Ball,
only two of whom had higher scores. Collishaw went on to command
3 Squadron RNAS, which became 203 Squadron Royal Air Force on 1 April
1918. However, while the navy had an interest in air combat, using Sopwith
*Pup*s as anti-airship fighters launched from ships, the real attraction was
surface strike, attacking ships and submarines.[13]

Surface strike

It was rapidly realized that if an aeroplane could drop a bomb, it could
also drop a torpedo. Not all torpedoes were suitable to be dropped from
an aircraft, and a torpedo was a very heavy weapon by the standards of
the time. The earliest torpedo bomber, the Short *184*, was only able to
carry a small 14-inch Mk X torpedo weighing 765 lbs. The *184* came into
service in 1915; over 900 were built and it remained in service until 1921.
It was to the Dardanelles that the Admiralty deployed the first aircraft into

Fig. 6.6. Short *184* dropping a torpedo. With increasingly powerful
engines, the Short *184* remained in service until 1921.

service aboard the seaplane carrier HMS *Ben-my-Chree*, and on 12 August
1915 two aircraft attacked and sank a Turkish ship. In fact one of the aircraft
had engine trouble and, having landed, taxied on the water to carry out its
attack. An entirely airborne attack, which successfully sank another ship,
was carried out five days later.[14]

When she arrived in the Eastern Mediterranean the *Ben-my-Chree* had
joined the seaplane carriers *Ark Royal*, *Empress*, *Anne* and *Raven II*. During the
Gallipoli Campaign their aircraft established Allied air supremacy in the theatre, undertaking reconnaissance and bombing as well as air defence against
German and Turkish aircraft. Once the Gallipoli Campaign ended, fighting
continued in the Aegean and the ships were placed under the command of
the general officer commanding in Egypt. These ships constituted what has
been described as a carrier task force, 'the first in the British Navy and the
only such specifically organised naval force anywhere during the war'.[15]

Anti-submarine

The Admiralty had been interested in using aircraft to attack submarines
since 1912, and to do so it developed large patrol aircraft for both maritime
reconnaissance and the anti-submarine role. Early in the war it ordered a

large prototype seaplane from the boat- and shipbuilder J. Samuel White & Co. It was intended for coastal patrol, with the range to bomb the German fleet in Kiel, and was termed the 'AD Seaplane Type 1000'. Initially the unreliability of engines meant that aircrew were unwilling to fly too far from land. As engines became more dependable and large long-endurance flying boats came into service, the use of aircraft against submarines blossomed. The leading light in developing anti-submarine aircraft was John Porte. He had been invalided out of the navy in 1911 with tuberculosis. Ironically, for the man who was to develop the best anti-submarine aircraft of World War I, he had been himself a submariner. He developed an interest in aviation and teamed up with the American designer Glenn Curtiss. When war came, he re-entered the navy in the RNAS. His influence led to the purchase of Curtiss *H4 America* flying boats. They proved insufficiently strong and were underpowered. Porte developed the *H4* with a new hull, which became the *Felixstowe F1* (named for the naval base at which Porte worked). He did the same for the Curtiss *H12*, which entered production as the *Felixstowe F2*. Despite being a very large aircraft for the time, the *F2* was surprisingly manoeuvrable; on one occasion five of them were attacked by 14 German seaplanes, of which six were shot down for the loss of one *F2*.

The navy also used airships. Unlike the Germans, the British did not invest heavily in rigid airships after the *Mayfly*, building only nine during the war. It did, however, use a lot of non-rigid airships – what are now called 'blimps'. A shaped gasbag of hydrogen had a car or 'gondola' slung under it. The first was the *SS*-type, which had the fuselage and engine (minus wings) of a *BE2* aircraft slung underneath as a combined crew and propulsion gondola. This was successful as a patrol aircraft and was succeeded by the larger *Coastal* type, which had a twin-engined Avro fuselage. The ultimate was the *SSZ*, which was the 'finest non-rigid in the world'; it had an endurance of 12 hours and, like the other airships and the large flying boats, carried a wireless as well as depth charges.[16]

It was recognized very early on that the aircraft did not need to sink a submarine, or even to see it. The threat or perceived threat of an attack was sufficient to make a submarine dive. German submarines could make as much as 16 knots or more on the surface, but only between 2 and 8 knots submerged (and that for no more than an hour), after which their

Fig. 6.7. The *Felixstowe F2* maritime patrol and anti-submarine aircraft. As well as depth charges to use against submarines, later in the war it carried a hydrophone to lower into the water to detect them, having alighted on the sea (see Fig. 7.4).

Fig. 6.8. The *SSZ* non-rigid airship. By the end of the war some were equipped with a hydrophone to lower into the water, some 40 years before helicopters began 'dunking' sonar.

batteries would be exhausted. Even at the most economical speed of 4 knots submerged, they could only then travel 60 miles. A surface ship was much faster than a submerged submarine, so attacking a surface ship was difficult unless a higher-speed surface approach was possible. Thus the mere presence of an aircraft around a convoy was often sufficient to protect it; by the war's end the standard practice was for a convoy to be accompanied, where possible, by two aircraft, non-rigid airships or flying boats: one to deter, the other to attack a U-boat.[17]

The RNAS was not satisfied with just defending; its patrols were intended to seek for submarines actively. This led to the famous 'spider web' search pattern. Centred on the North Hinder light vessel at the eastern end of the English Channel, this was a very sophisticated search system and was named for its appearance on a chart; it utilized two aircraft patrolling abeam of each other at the limits of their visibility, thus covering an enormous area of sea rapidly. It was heavily reliant on signals intelligence from Room 40.[18]

Strategic bombing

The navy also developed bombers to attack land targets and surface ships in harbour. Initially these were intended to bomb the German fleet in harbour and the German U-boat bases in Flanders, but later they were used for out-and-out strategic bombing aimed at disrupting German industry.

Shortly after the outbreak of war the Admiralty Director of the Air Department, Captain Murray Sueter, approached the aircraft designer Frederick Handley Page. He sought what he called 'a bloody paralyser'. In response, Handley Page produced the first British strategic bomber, the *O/100*. This was a twin-engined biplane carrying up to 2,000 lbs of bombs, an immense bomb load for the time. They were deployed to France as the RNAS 7A Squadron in 1916. The RNAS planned to undertake not only what is now termed 'offensive counter-air' to attack the Zeppelins and submarines at source, but also strategic bombing. The Naval Intelligence Division had identified that the Germans were short of steel, and the steel factories of the industrial Ruhr valley were identified as targets. In 1916 attacks were launched against chemical factories, blast furnaces and

Fig. 6.9. Captain Sueter's 'bloody paralyser', the Handley Page *0/100* strategic bomber, seen here in France. Note the disruptive pattern camouflage, which was not unlike that being used on ships.

munitions factories in Germany. Unfortunately, while the French were welcoming, the British army was very much against the navy running a bombing campaign from France. The *0/100*s were returned to England after an inter-service dispute that went all the way to the Cabinet.

The bomber was developed into the *0/400* by substitution of more powerful engines and returned to France in 1918 as part of the Royal Air Force. They undertook strategic bombing, and apparently now the army did not object. In the meantime the navy remained wedded to the idea of strike from the sea, seeing it as being more versatile than from fixed land bases.

The birth of the aircraft carrier

It will have been obvious from the foregoing that the Zeppelin was seen as a major threat from the start of the hostilities. Being able to launch aircraft closer to the enemy bases would obviously confer great advantages. The RNAS had successfully launched a fighter aircraft, in this case a Bristol *Scout*, from an airborne flying boat on 17 May 1916, but this technique of an aerial aircraft carrier was not pursued. The navy instead saw launching aircraft from ships as being the way ahead. Initially, ships carried seaplanes to be lowered by crane into the water to take off and to be recovered after landing, which is what *Vindex*, accompanying the BCF at the battle of Jutland, had to do. Her aircraft saw and reported the German fleet, but unfortunately the reports were not received by Admiral Beatty.[19]

It was obvious that using a wheeled land plane rather than a seaplane was the way ahead; lowering a seaplane into the sea for it to take off just took too long. The next step was to fit *Campania* with a longer flight deck to allow seaplanes on a detachable wheeled trolley to take off from the ship. The trolley would fall away once the aircraft was airborne, but they still had to land on the sea and be lifted back on board with a crane. It was obvious that what was needed was a ship that could launch wheeled aircraft rather than floatplanes and allow them to land back on the ship. In fact on 3 November 1915 *Vindex* had already launched a land plane from a short flight deck on her bows. On 2 August 1916 another *Scout* launched from *Vindex* unsuccessfully attacked a Zeppelin. The problem was that while the *Vindex*, *Campania* and the other seaplane carriers had given excellent service, they were converted merchant ships and, while they could keep up with the fleet, they were rather small.[20]

While he was First Sea Lord during World War I, Fisher conceived his 'Baltic Project'. In essence, he wanted to land a Russian army on the German Baltic coast. The project never came anywhere near to fruition; however, three specialist 'large light cruisers', HMS *Furious*, HMS *Glorious* and HMS *Courageous*, were built as part of it. All were ultimately to be converted into aircraft carriers, but the last two were completed in their original form. *Furious* was converted while still being built. Instead of one of her 18-inch guns, she was completed with a large hangar forward of the superstructure, with a 160-foot-long flight deck on top of it. In August 1917 Squadron Commander Edwin Harris Dunning landed a Sopwith *Pup* onto this deck, the first time this had ever been done. Unfortunately he was killed trying to repeat the feat. It was apparent that trying to fly round the superstructure was extremely difficult, so *Furious* then had her after gun turret removed and a second 'landing on' deck built over the stern, leaving a funnel and superstructure in the middle of the ship. That this was done in four months at the end of 1917, when there was a need to build ships and escorts because of the German campaign of unrestricted submarine warfare, is some indication of the priority that this was given.[21]

Fig. 6.10. HMS *Furious* with both forward and aft flight decks. She has
an *SSZ* non-rigid moored to the aft deck. The vertical structures around
the forward flight deck are to break up the wind flow over the deck.

Unfortunately the new deck was not satisfactory; only three attempted
landings by Sopwith *Pups* out of 13 were successful, and the deck was
deemed 'unusable, even for wartime service',[22] when greater risks
could be accepted. *Furious* was converted after the war to have an
unobstructed flight deck. In the meantime the HMS *Argus* was being
built. She was the 'first flush-deck aircraft carrier capable of operating
wheeled aircraft to complete in any navy'. Unfortunately the navy was
about to lose control of its aircraft, and with them the Royal Naval
Air Service.[23]

The birth of the Royal Air Force

It is curious that those who were involved in the decision to merge the Royal Flying Corps and the Royal Naval Air Service into a single new service wrote so little about the reasons for their decisions. Even now it is difficult to understand fully the machinations that led to the formation first of an Air Board and then of an Air Ministry and a new service. Ultimately it was a political decision; it was not driven by military need or pressure. The initial cause appears to have been an inter-service dispute over aircraft engines. At the start of the war, Britain was reliant on France as a source of aero-engines, but it rapidly started building French engines under license and then developed indigenous engines. The Royal Aircraft Factory tended to use foreign-derived engines, such as the Hispano-Suiza in the *SE5a*, whereas the Admiralty, despite having been the original British 'sponsor' of the Hispano-Suiza, used British-derived engines, most notably the Rolls-Royce Eagle in the Handley Page *O/100* and *O/400* (later the American Liberty engine was widely used). The Air Board was formed in January 1917 to resolve disputes and to prioritize research and manufacture. Despite this, disputes continued; as already mentioned, the naval deployment of strategic bombers to France without army agreement went to the Cabinet for resolution. At the same time, responsibility for the air defence of the United Kingdom had been passed back to the RFC; meanwhile, the Germans continued with Zeppelin raids on London, which were followed in 1917 by *Gotha* bombers operating out of bases in Belgium. In reality the raids were pinpricks compared with what was to follow in World War II, but the shock of the new, combined with an apparent inability of the two established air arms to prevent or initially to defend against the attacks, made for a public outcry that 'something had to be done'. Unfortunately the matter now came down in large part to personalities. The new Air Board, intended to resolve inter-service disputes, was headed by Lord Curzon. He wrote a report to the War Committee, a sub-committee of the Cabinet. It 'comprised a harsh, even strident attack on the Admiralty's organisation', including 'the alleged failure of the Admiralty to develop rigid airships'. Why he believed that they should have done so, Lord Curzon did not explain.[24]

Lloyd George, the Prime Minister, was not known as the 'Welsh wizard' for nothing. His government was being criticized for the lack of home air

defence, and to meet the criticism he produced General Smuts. Jan Smuts
was a South African who had fought against the British in the Boer War
and for the British as head of the South African army in East Africa during
the early part of World War I. Lloyd George had invited him to come to
London in 1917 to join the Imperial War Cabinet and then to report on
the way ahead for military use of aircraft. Smuts's attitude was that the two
services did not grasp the new warfare arising from the aeroplane, and the
best answer was to form a new service. One author has even gone so far
as to suggest that Smuts was taking revenge for his defeat at the hands of
British forces in the Boer War by making a recommendation 'that would
seriously damage Britain's seapower'. That is fanciful. Lloyd George was
a consummate politician who knew how to pack a committee to get the
answer he wanted, and he wanted a dramatic gesture to head off public
opinion over the German bombing raids on London.[25]

When the Smuts report was circulated for comment, the First Sea
Lord, now Jellicoe, was adamantly against the navy losing the RNAS, as
was the First Lord, Geddes. Unfortunately Jellicoe's star was waning, and
when Beatty's opinion was sought he apparently believed that all it took
to become a naval aviator was the ability to fly, and that there was no need
for the navy to retain an independent air arm. 'His seemingly supine attitude [...] is beyond rational explanation.' Beatty's opinion was enough to
carry the debate, and the navy lost the RNAS. Beatty then spent his years
as First Sea Lord unsuccessfully attempting to regain control of naval air,
which did not actually happen until 1937, and then only of carrier-borne
aircraft, not shore-based maritime patrol assets.[26]

In his war memoirs Lloyd George was far more concerned with the
machinations concerning the appointment of an Air Minister in October
1917. Once that had been done, the Royal Air Force was created under the
future Marshal of the Royal Air Force, Lord Trenchard, who actually initially
did not even want the job, not believing that the RFC should be split away
from the army. The naval aviators were given the option of transferring
to the new service; most did so, so the navy was to be robbed of aviation
expertise and advice for many years. Most of what the new service gained
in terms of technology that would allow it to start long-range bombing
came from the RNAS; this included not only navigation instruments, but
also direction-finding wireless telegraphy and a bombsight that the RAF

continued to use for two years into World War II. However, there was a certain inertia before the Royal Air Force fully came into being, and the navy continued until the end of the war with its ambitious plans for surface strikes.[27]

The Tondern raid and beyond

Following the first attack in 1914, the Grand Fleet had repeatedly attempted to attack German bases, mostly Zeppelin sheds, with sea-based aircraft. The seaplane carriers were usually supported by the Grand Fleet, but did not succeed in provoking a response from the *Hochseeflotte*. Despite no longer having control of the Naval Air Arm, the navy continued to press ahead with attacks on German Zeppelins in their bases. An airship base at Tondern (present-day Tønder in Denmark) was identified as a good target. Initially planned as Operation *F6*, the attack was delayed by bad weather. On 17 July 1918, *Furious* launched the first carrier strike in history – that is, carried out by wheeled aircraft from a flight deck rather than by seaplanes. Seven Sopwith *Camel F2*s took off from her forward flight deck to attack Tondern. They destroyed two Zeppelin airships and a captive balloon in their sheds. Two aircraft landed in Denmark, while four returned to the ships (one prematurely because of engine problems) and landed in the sea for their pilots to be recovered.[28]

Following the success of the Tondern raid, a far more ambitious strike was planned on the *Hochseeflotte* itself. On 10 October 1918, 185 Squadron RAF embarked in *Argus*, equipped with Sopwith *Cuckoo* aircraft that could each carry an 18-inch Mk VIII torpedo. They commenced training for an attack on the *Hochseeflotte* in Wilhelmshaven, but the end of the war meant that the attack did not take place. The Fleet Air Arm had to wait almost exactly 22 years until it could attack an enemy fleet in harbour at Taranto.[29]

Naval air

The RNAS is now recognized as having been a powerhouse of thinking about the applications and use of air power during its short existence. It is unfortunate that it was subsumed into a much larger service, depriving the navy of its air arm for a vital 20 years. Thus what became the Fleet Air Arm entered World War II with the most modern and advanced aircraft carriers in the world, procured by the Admiralty, and some of the most antiquated aircraft in service, provided by the Royal Air Force. Despite this, it was to perform extremely well during the latter war.

As well as embracing above-surface warfare, the Royal Navy took to underwater warfare and made major strides in this field as well.

The submarine

A recurring theme of this book is the historical misperception that the Royal Navy before World War I was hidebound, arrogant and above all resistant to change. As in any large organization, there were elements whose writings and behaviour can be used to support such a view. Submarines were held to be 'a damned un-English weapon'. The man who wrote that, Admiral Sir Arthur Wilson, was the man who first suggested that the Royal Navy should purchase submarines, because the French were doing so. He didn't like the idea of submarines, but recognized that they were likely to be vitally necessary. The repetition of such selective quotations has obscured the actuality that the Royal Navy was very much the leader in developing the submarine into what would now be called a weapons system, and it thought long and hard about how to use it: what is now called operational doctrine. The navy also looked at and experimented with various different submarine concepts, including submarine monitors, anti-submarine submarines and high-speed steam-powered submarines.[30]

There had been numerous experiments over the centuries to make a submarine. Throughout, the Royal Navy had maintained a studious apparent disinterest. It in fact watched developments closely, particularly those in the United States and France. As early as 1886 it had established a committee tasked with assessing developments in submarine technology. In that year there was even an unfortunate trial of a 'submarine boat', in which two captains and Sir William White, the Director of Naval Construction, participated. While the trial was unsuccessful – having dived, the submarine had much difficulty in coming back to the surface – fortunately no one was lost. Only very late in the nineteenth century, when the Admiralty started receiving reports of working French submarines, did it see the need to act, because, before the naval rise of Germany, the French were still seen as the most likely enemy. Their possessing a weapon that combined the also new locomotive torpedo was an obvious threat. The Admiralty wanted to see how practical such vessels were and also to evolve defences against them. Accordingly, in 1901 they appointed Captain Reginald Bacon as Inspecting Captain of Submarine Boats. He was then to be the driving force behind the development of the submarine, although Admiral Fisher was also an early enthusiast.[31]

As a first step, the Admiralty decided to build an American submarine under license. Ironically, these were the brainchild of John Holland, an Irish-American, whose major aim in life was to destroy British naval supremacy as a way towards Irish independence. Holland's boats were small, but they combined technologies that made them into a viable submarine: the internal combustion engine dependent on air for propulsion on the surface, electric underwater propulsion from batteries charged by the engine while surfaced, and the torpedo. The navy added to the basic design a periscope to allow it to attack a target when submerged. They were not very seaworthy; when they were on the surface their freeboard (the amount of vessel above the waterline) was minimal, making them vulnerable to being swamped. Once there were *Holland*-class boats in service, Bacon instituted a series of trials. Supposedly these were to evolve defences against submarines, but Bacon rapidly used them to develop submarine tactics for use against surface ships. Despite obvious weaknesses, including minimal armament, a single torpedo tube and a short operating range, they proved the concept.[32]

Even while the five *Holland*-class boats were completing, the Admiralty had ordered a larger submarine, the *A*-class. To improve their seaworthiness, they had a conning tower. They were enlarged after four had been built, and the last, *A13*, had a heavy oil engine rather than a petrol one to reduce the risk of fire. At this stage heavy oil or diesel engines were not mechanically reliable, so despite their attractions the Admiralty persisted with petrol for a few years more.

Successor classes followed, each larger and more advanced, with larger and increasing numbers of torpedo tubes. By 1913 the navy was bringing into service the *E*-class. At the time this was the best submarine in the world. It had five torpedo tubes, some had a deck gun, and it was diesel-engined. The class was to perform extremely well during World War I.

Fig. 6.11. A-*class Submarines in Portsmouth Harbour* by H.L. Wylie.

The Admiralty had spent a lot of time and effort thinking about how submarines should be employed. What is now seen as the obvious use (attacking merchant shipping covertly) was seen then by the Admiralty and the wider governmental authorities as unthinkable. If submarines were to be used for a war against trade, it would have to be in accord with the usages of war. The early German abrogation of these rules and unrestricted submarine warfare is discussed in the next chapter. It did not feature in pre-war British planning; submarines were warships and would be used against other warships. Indeed the Royal Navy conducted submarine warfare in accordance with international treaty throughout the war, even if this meant exposing themselves to great risk, such as when *E2* operated in the Sea of Marmara during the Gallipoli Campaign, ensuring the safety of the crews of merchant vessels before sinking them.[33]

Initially, as they were short-ranged and not suitable for long periods at sea, they were seen as a defensive weapon to protect ports and anchorages. As their capabilities improved it was realized that the early visionaries who included Admiral Fisher had been right; submarines were an ideal weapon, and not just locally. He evolved the concept of what has been termed 'flotilla defence'. In his vision, the maritime defence of Britain would depend on 'swarms' of torpedo-armed surface and submarine vessels. The idea did not survive the end of Fisher's first period as First Sea Lord. Now

Fig. 6.12. An *E*-class submarine, the first of which entered service in 1913.
Compare this with a contemporary German submarine in Fig. 1.4.

the emphasis was on using submarines, controlled from a surface ship, to attack the enemy's battlefleet; indeed it was on that basis that submarines were present at the battle of Heligoland Bight.[34]

Personnel

Submarine service required a somewhat different type of sailor to the rest of the fleet, one able to live in very confined quarters and subject to far greater risks. Why the submarine service is known to this day as 'the trade' is unknown, yet it predates World War I. The first use of the term appears to be have been by Kipling, writing in 1914, but he uses it as if it were well established when he wrote: 'They play their grisly blindfold games / In little boxes made of tin. [...] That is the custom of "The Trade".' Certainly in the Edwardian era being 'in trade' was still a pejorative term, implying a degree of social unacceptability. It probably arose because officers serving in submarines had necessarily to get their hands (and a lot else besides) dirty; submarines were a cramped, uncomfortable, insanitary and above all dangerous environment; significant numbers of submarines were lost in accidents. As is the nature of such terms, 'the trade' became a badge of pride. The other great pride is that service in submarines was voluntary, for both officers and ratings. What was probably an incentive was submarine pay, which was substantial. For example, a chief petty officer whose basic pay was 2s. 8d. per day would also receive 2s. 6d. submarine pay.[35]

Unlike the other branches, the submarine service had far fewer opportunities and appointments for officers as they became more senior; indeed there were few appointments for lieutenant commanders, and most of those were in command or as staff officers. The result was that significant numbers of lieutenants and lieutenant commanders returned to general service, most usually because they were either not recommended for submarine command or because, having been recommended, they then failed the qualifying course for command, latterly known as 'the perisher', which was established in 1917.[36]

Submarine service was popular because its very nature encouraged *esprit de corps* and allowed and even required initiative in a way unthinkable in the major units of the Grand Fleet. That the operational efficiency of the submarine service was maintained throughout the war despite a fourfold increase in personnel (from 1,418 to 6,058) reflects great credit on the organization. A lot of the personnel who entered the submarine service came from the Royal Naval Reserve. Particularly sought after were those qualified to be navigating officers, which meant they had to have a Board of Trade certificate as a mate at least. In the days of GPS and other modern aids to navigation, it tends to be forgotten how difficult navigation used to be. In an earlier chapter, reference was made to errors of eight miles being made in the course of a day's steaming by a surface warship. This was by no means regarded as unusual. Navigating a submarine was much more difficult still, even on the surface. The very low height of the bridge above the surface meant that the horizon was very close, and so making observations of marks and land features was difficult even in calm weather. To compound matters, many peacetime navigation marks such as lightships and buoys were removed at the outbreak of war. When submerged with only a periscope to make such observations, navigation became a matter of 'by guess and by God'. Not for nothing did the navy look for skilled navigators for submarines.[37]

The dangerous nature of service in submarines meant that imperturbability and coolness under pressure became the 'norm' in the submarine service. One episode that became a legend within the submarine service concerned the future Admiral Horton (then a lieutenant commander) in command of *E9*. His submarine was bottomed in the Heligoland Bight while being searched for by the Germans with sweep wires. With nothing to do but wait, the officers played bridge in the control room, which would have been in full view of most of the ship's company. When one officer revoked, distracted by a sweep wire noisily catching on the bow of the submarine, Horton told him that he had to concentrate if he was to play bridge well. Such coolness was to be required throughout the war.[38]

The war

As tensions increased during the summer, ten submarines of the 8th Submarine Flotilla under Commodore Keyes deployed with its depot ships to Harwich. He planned, in view of the 'possibility of the German Fleet proceeding to the southward to attack the Expeditionary Force in transit', to maintain four submarines on station to attack them. As early as October 1914 three submarines were dispatched around the north of Denmark into the Baltic to work with the Russians against the Germans who used the Baltic as a training ground. One of the submarines was unsuccessful and had to return to Britain, but the other two remained in Russia, under Russian operational control. Gradually this force was increased and had some notable successes against German warships, notably *E8* in August 1915 when she sank a German cruiser. They also conducted a successful war against merchant shipping, but abided by the usages of war, that is surfacing and ordering the ship to stop, and even in one case putting on board a prize crew rather than sinking the ship after determining it carried contraband. Their activities restricted not only German imports of iron, but also the sailings of warships from Kiel to the extent that 'even the trial trips of some newly launched ships [...] had to be postponed'.[39]

The submarines also maintained patrols in the Heligoland Bight, and senior officers of the German fleet appear to have caught the disease that afflicted their British counterparts at the beginning of the war, 'periscopitis'. This was not surprising. One submarine, *H5*, actually left her assigned patrol area and entered the mouth of the Weser River, where Lieutenant Varley attacked and sank a U-boat. He surfaced and attempted to make a German survivor prisoner, but was forced to dive by understandably angry German destroyers. Despite disobeying his patrol orders, he was awarded a DSO for sinking the submarine.

Overall, British submarines were used effectively while operating within the constraints of international law. Certainly if they had operated as the German submarines did, the Baltic submarines could probably have substantially reduced German steel imports from Sweden.[40] Britain did not embark on unrestricted submarine warfare in the way that the Germans did, and so did not sink merchant ships without warning. Even if she had, however, since the German merchant marine had largely ceased to exist at the outbreak of war, there would have been scanty merchant ships as targets. As with the Grand Fleet, the submarine service was to spend large parts of the war waiting for targets; the German navy, apart from its own submarines, spent little time at sea. Nonetheless, the submarine service was a constant threat to the German navy throughout the war.

Technical developments

Unfortunately, little documentation has survived about the thinking behind technical developments during the war or the expansion of the submarine service. However, the thinking can be surmised from the submarines produced, of which there was a great variety. Submarine classes were alphabetically designated and by the war's end the *F, G, H, J, K, L, M, N, P* and *R* classes had all entered service, suggesting that the submarines comprised a productive arm of the navy. What follows is intended solely to give a broad picture of developments.[41]

There was a continual hankering after a submarine that could operate with the surface fleet, a submersible destroyer that could attack the enemy battle line. This required a submarine capable of a surface speed of 21 knots or better. The fastest diesel-powered submarine was the *J* class, at 19 knots. So the Admiralty went for steam; after all, they were familiar with steam plant. While the concept of a fleet submarine was a wrong turning, the resulting *K* class were a significant technological achievement; with eight torpedo tubes, two guns and the capability of making 24 knots on the surface, they were theoretically a formidable weapon. However, a series of accidents culminated in the infamous battle of May Island on the night of 31 January to 1 February 1918. In a tragic farce resulting from an attempt to operate submarines with surface ships, there were five collisions resulting in the loss of two *K* class and many of their crew. This finally proved that the concept of a fleet submarine was flawed; surface ships and submarines could not work together.[42]

Mention has been made earlier of the use of monitors for shore bombardment and their vulnerability to enemy shore batteries firing back. Using a *K*-class hull but with diesel engines, the *M* class carried a 12-inch gun. It was intended that the gun would be fired while aiming through the periscope, with the gun muzzle protruding above the surface. It worked astonishingly well, but only *M1* was completed during the war.

Another innovative design, albeit in a totally different direction, was the *R* class, which was a classic example of an idea before its time. These were intended as U-boat hunters; they were small but heavily armed, with six torpedo tubes in a very streamlined hull. Unlike any other submarine for another 30 years, they were much faster submerged than on the surface. Their weakness was that they had no accurate means of detecting submarines while submerged, apart from a periscope. They carried an extensive array of hydrophones and are believed to have had a means of processing the information received. However, the hydrophones were just not accurate enough. An anti-submarine submarine needed a proper sensor, like sonar, which then didn't exist. The *R* class was an extremely advanced submarine, but ahead of the technology it needed to be effective in its designed role.[43]

The navy and new technologies

In the period before and during World War I the Royal Navy did not just accept new technologies, it actively embraced them and pushed them to their limits. This chapter has looked at some of them, but it must not be forgotten that the navy had also pioneered many other technologies, including the use of wireless, wireless direction-finding and signals intelligence. It had developed the hydrophone for anti-submarine use and developed a range of anti-submarine weapons. As this chapter has shown, on land it can fairly claim to have originated the armoured car and, if not to have fathered the tank, then certainly to have been at the very least a benign uncle. In the air it was the major driving force behind military aviation until its air arm was taken away for political and not military reasons. At sea it led the way in the development and use of submarines.

CHAPTER 7

The British and German Wars against Trade

...that we may be a safeguard unto our most gracious Sovereign Lord,
King GEORGE, and his Dominions, and a security for such as pass on
the seas upon their lawful occasions...

'Prayer for the Royal Navy' (extract)

Seapower is ultimately a means of influencing events on land, particularly trade, as the prayer for the Royal Navy quoted above implicitly recognizes. It was the maritime trade war, and not a clash of battlefleets, that ultimately decided the outcome of World War I.

Globalization is commonly thought of as being a late-twentieth-century phenomenon. While the term was, the reality was not. Before World War I the nations of the world were extremely interdependent economically. The degree of globalization achieved by 1914 was not to be matched again until the 1990s. Many felt that the degree of interdependence made war impossible between the major powers.[1] While harder-nosed thinkers realized that war was not only possible but increasingly likely, some politicians and in most countries the military paid little attention to the mechanisms of trade and even less to economics. The British Admiralty was the major exception.

At the beginning of the twentieth century, having led the way towards industrialization, Britain's industrial pre-eminence was slipping as other nations caught up. London, however, remained the centre of world banking and was the financial powerhouse for world trade. Just as important, the British Merchant Navy remained by far the largest in the world. In 1915, despite war losses, it was comprised of 3,063 ships out of a worldwide total of 4,881.[2] The Admiralty recognized these facts and also, vitally, that the whole country was fed and manufacturing industries supplied on the basis of a 'just in time' supply chain. Having industrialized and moved on from being an agricultural society, Britain was far from being self-sufficient in food. At any given time, there was sufficient food in the country for three weeks' consumption; the country's granaries and food warehouses were literally at sea. This made Britain very vulnerable to any interruption of world trade.

Germany was similarly vulnerable because it imported about 25 per cent of food for its people, who annually consumed 1.5 million tons of wheat from America, and for its farm animals, which annually consumed 3 million tons of barley from Russia – half of Germany's yearly requirement. Because German soil tended to be poor, its agriculture was very dependent on imported fertilizers. As for raw materials, the United States supplied all of Germany's cotton and 60 per cent of its copper.[3]

Blockade and international law

Historically, before air freight, all international trade was conducted by sea. Recognizing this, Britain had traditionally used economic warfare and blockades of its enemies' trade as the preferred way of fighting a war. This entailed stationing warships outside an enemy's ports and preventing warships and merchant ships from entering or leaving, usually by capturing them as 'prizes', which was described as a 'close blockade'. As early as 1903 the Naval Intelligence Division, which was in everything but name the Naval War Staff, had said that in the event of a war with Germany, geography gave Britain an enormous advantage, sitting as it did 'like a breakwater, 600 miles long [across] the path of German Trade with the west'. However, by the first decade of the twentieth century it became obvious that a close blockade would no longer be militarily viable. Not only would it be no longer possible to confine enemy warships to their bases, but merchant ships would also have freer movement. The advent of the torpedo, the submarine and the sea mine made it impossible.[4]

International law

Despite the opinion of General Sherman, who had said, 'War is cruelty. There is no use trying to reform it', as well as similar sentiments expressed by others, during the nineteenth and early twentieth centuries there were many attempts to regulate and codify maritime warfare, particularly as it affected trade. Looking back at the events of the twentieth century, it can be difficult to realize that prior to World War I most nations appeared genuinely to believe that international agreements were binding and would be adhered to.

The Paris Declaration of 1856 was the first attempt to set out an internationally agreed framework for the conduct of war at sea. The conference had initially been convened with the intention of abolishing privateering. Privateers were privately owned armed vessels licensed by a government to attack the trading vessels of an enemy nation. The United States had relied heavily on privateers in the war of 1812, so it was hardly a surprise that the United States was not a signatory.[5]

The conference and the subsequent declaration not only defined what constituted a naval blockade of merchant shipping, but stated that a neutral flag 'covers enemy's goods, with exception of contraband of war', which meant that for contraband, goods carried in a neutral ship were not liable to capture by a combatant. Of the other major nations ultimately involved in World War I, all were signatories (Germany signed as Prussia).

The Hague Convention of 1907 set out to go much further. Unfortunately Treaty XII, which would have established an international prize court to arbitrate on disputes, was never ratified. Thus when Britain established its blockade of Germany in 1914, it had to rely on its own prize court to arbitrate on its own actions.

What was to cause Britain much heart-searching was the London Declaration of 1909 (often referred to as 'The Prize Regulations'), which arose out of an international conference in London the previous year. Despite being a signatory, the British government, along with every other signatory, never ratified the treaty. At the beginning of the war Britain announced that it would consider itself bound by it. This was despite the fact that it realized it could not implement a close blockade, and would have to use a distant blockade, which Chapter I, Article 1 expressly forbade. There were other provisions that would further work to the disadvantage of the blockading power, such as Chapter I, Article 19, which precluded the capture of a vessel on passage to a neutral port, even if it was carrying cargo ultimately destined for a combatant.

Germany made much use of this provision. Despite the restrictions on her own trade, Germany was actually to receive large volumes of re-exports, especially from the Netherlands. However, Britain did have one great advantage: the size of its merchant fleet meant that it carried a significant portion of world trade, because there was insufficient capacity under other flags. This meant Britain could actually decide which cargoes would go where. In peacetime this was not a problem, as British ships wanted the business; in wartime, however, it was to be a means to control maritime trade.

Chapter XI of the London Declaration defined what was 'absolute contraband' and thus liable for seizure: armaments, explosives, etc. It defined as 'conditional contraband' cargoes such as animal feedstuffs if they were destined for the use of a combatant's armed forces. There was also an extensive list of materials that could not be declared contraband. The last included many items that would now be termed 'dual use' (i.e. materials that were intended for civilian purposes but also had a military application). This included anything from cotton (for explosives and uniforms) to fertilizers (for foodstuffs and the manufacture of explosives). This and the attitude of the United States led to what was at times a 'vacillating policy' from the British government as it tried to stop Germany being supplied without upsetting her suppliers.[6]

When it came to instituting a blockade of Britain, Germany was in a much weaker position, although its ready access to the Baltic made for ease of maritime trade with Scandinavian nations, particularly Sweden, a friendly neutral. Otherwise its access to the sea was extremely constrained. It had, of course, ports on the North Sea, but the exits from the North Sea into the Atlantic Ocean were controlled by the Royal Navy. Apart from some use of surface raiders described earlier and some forays by elements of the *Hochseeflotte* against Norwegian traffic, Germany's main blockading effort was through submarines.

Planning for war and the 'July crisis'

The Royal Navy had an extremely sophisticated war plan based on Germany's being the likely enemy, and based on maritime economic warfare:

Blockade was central to the British Government's war fighting. The Committee on Imperial Defence, during prolonged deliberations before the outbreak of war, had identified Germany's food supply as her 'Achilles' heel'. Thus, as part of a policy of 'bringing the utmost possible economic pressure to bear on the enemy' it was intended to systematically attack her food supplies.[7]

The Admiralty expended a lot of effort on considering how economic warfare would be conducted in the future. In 1908 it had produced a document that was considered by the Committee on Imperial Defence, the highest politico-military body in the British Empire. It was normally chaired by the British Prime Minister. The CID accepted the case for economic warfare, but this posed a problem for the Admiralty: the necessity of a distant blockade. As a blockade, with warning, could to some extent be circumvented, the Admiralty's war plan was not widely promulgated. This was read by some as meaning that the Admiralty did not have a war plan. In fact its very detailed plan emphasized the need for speedy action at the outbreak of war.

In July 1914 the international commercial and banking worlds began to realize that a major war was likely; this became known as the 'July crisis'. Within days, nearly every stock market in the world had closed and interest rates started rising rapidly. In Britain they went from 3 per cent to 10 per cent and the pound sterling, a gold-backed currency, rose to the extent that the 'world foreign exchange system effectively shut down'. At the same time maritime insurance, which covered both ships and their cargo, became extremely expensive: premiums were as high as 20 per cent of the value of the cargo.[8]

Rising insurance rates were a major cause for concern. A significant factor in the defeat of the United States in the war of 1812 had been insurance. Towards the end of that war no one, not even American companies, would insure American merchant ships or their cargo. Owners would not put their ships to sea uninsured, and merchants would not entrust an uninsured cargo to an uninsured ship. That war ended with the American merchant fleet confined to harbour, which caused grave economic hardship in the United States along the whole of its eastern seaboard. It was not surprising that, having inflicted this defeat on the Americans, almost exactly a century later the British were extremely aware of the importance

of insurance for ships and their cargo. By the end of July 1914, even before
war was declared, 'the market for war insurance was already drying up'.
A scheme of state-backed insurance already prepared with oversight from
the Committee on Imperial Defence was implemented on 4 August and
subsequently copied by other belligerents and neutral nations.[9]

Britain's economic war on Germany

Before the war broke out, the Admiralty began implementing Britain's
economic war on Germany. It first instituted a prohibition on the export
of animal and animal feedstuffs and coal, and then at the outbreak of war
it instituted a distant blockade at either end of the British 'breakwater'. As
part of the war book's procedures, on 4 August a royal proclamation was
issued which specified what Britain regarded as absolute contraband and
conditional contraband. On 20 August a further proclamation ordered that
'the Commissioners for executing the office of Lord High Admiral' were
to set up a prize court to adjudge whether ships and cargoes captured by
blockading vessels were indeed contraband.[10]

   Germany had anticipated that Britain would institute a blockade, but
had planned for a short war. It expected that France would be defeated
by a rapid invasion through Belgium, into France and encircling Paris.
Germany anticipated that Russia would be slow to mobilize and that while
France was being defeated, a relatively small blocking force would suffice
in the east. With France defeated, Germany would then turn on Russia.
Nonetheless, Germany had done some planning for imports through neutral
countries, particularly the Netherlands, as cargoes could then be shipped
up the Rhine from Rotterdam by enormous barges known as *Rheinschiffe*.[11]

   To complicate Germany's position, most of her merchant marine, which
at the start of the war comprised about 12 per cent of the world fleet, was
either captured by the Allies or confined to neutral ports. However, with
the active connivance of merchants in the United States and other neutral
countries, the Germans found ways of circumventing the restrictions placed
on their merchant marine. Many German companies established branches in
neutral countries; for example, the famous Hamburg America Line became
the Danish-based American Exporters Line. The German branch of the
Standard Oil Company sold its tankers to the American parent by using a

backdated contract. By backdating the contract(s) so that they appeared to have been made before the outbreak of war, it circumvented – and broke – international laws.[12]

Within two weeks, government policy started to vacillate. Mindful of the effects on the neutral countries bordering Germany, the Cabinet started reducing the range of the goods defined as contraband. Thus, almost from the outset, Britain, having instituted a blockade of Germany, was to be hampered in its implementation by having to deal with neutral countries. Neighbouring countries provided a conduit for German trade in both directions, as they continued to trade with the rest of the world. Britain in particular had to deal carefully with American ships and trade, as she was keen to avoid antagonizing a major trading partner that was to provide much of its munitions and was a potential ally.

If the connivance with breaking the blockade had been limited to the Germans and neutrals, the British government would have had an easier time of it. It has been argued:

> [T]he greatest untold scandal of World War I – though admittedly the case is largely circumstantial – was the degree to which contraband trade through neutral countries was financed by the City of London and carried across the Atlantic in British ships.[13]

While the neutrals' indirect trade with Germany via neighbouring neutral ports increased many times over, Britain's greatest problem was with the United States. The United States was a major seapower and had always insisted on its own belligerent rights in a time of war. It objected to Britain's attempting to stop its trade with Germany. This was in part commercial, but also because of internal political pressures; within the United States there was a large German element made up of recent immigrants who were quite naturally sympathetic towards Germany. Thus the blockade by both sides proceeded almost as a series of tit-for-tat actions. Germany declared a submarine blockade of Britain on 17 February 1915. The British government responded with a series of Orders in Council, which successively eroded the London Declaration. Its stepwise tightening of the blockade on Germany was aided, in the words of Maurice Hankey, the Cabinet Secretary, because 'it was Germany's misdeeds at sea that gave us the pretext gradually to free ourselves from the trammels of such instruments as the Declaration of Paris and the Declaration of London.'[14] Hankey was referring to such episodes as the sinking of the liner *Lusitania* on 7 May 1915, in which many US citizens were killed. Indeed the American ambassador stated that the London declaration 'would probably have given victory to Germany if the allies had adopted it'.[15]

The Northern Patrol

Germany's access to the sea and thus to maritime trade was threefold: the Baltic, the English Channel and the northern route (essentially between Norway and Britain). In the Baltic, she shared a sea-coast with her adversary Russia, but needed to trade with Sweden, particularly as the latter was a source of iron ore. Being narrow and easily patrolled, the English Channel was effectively closed to trade. This left the northern route as the only realistic way for Germany and the littoral neutral countries to continue their maritime trade, and it was here that Britain and her allies expended most effort in controlling and blockading trade.

The Royal Navy rapidly implemented the northern blockade as part of the measures resulting from the war book. This was to be undertaken by the 10th Cruiser Squadron (10CS). At the outbreak of war, Rear Admiral de Chair was admiral of the training service and commanded 10CS. The squadron was warned of the impending mobilization on 28 July, and three days later it was ordered to institute the northern blockade.[16]

The squadron was made up of eight elderly, indeed obsolescent, cruisers and armoured cruisers of the *Edgar* and the *Crescent* classes. The youngest, HMS *Crescent*, Admiral de Chair's flagship, had been launched in 1892. Among *Crescent*'s crew were eight cadets, the oldest aged 17, who were rated up as acting midshipmen and employed as junior officers. While the ships were of 7,000–8,000 tons displacement, they had been designed when the Royal Navy's emphasis had been the Mediterranean rather than Germany and the North Sea; they had not been designed for winter in the North Atlantic.[17]

The ships set off from their various home ports, and while en route up the west coast of Scotland to join the Grand Fleet at Scapa Flow came upon a German tramp steamer, the *Wilhelm Behrens*, on 5 August. She only

Fig. 7.1. The flagship of the 10th Cruiser Squadron
at the outbreak of war, the HMS *Crescent*.

stopped after HMS *Grafton* put a shot across her bows, and she was sent
into the Clyde with a prize crew on board.[18]

One of 10CS ships, HMS *Hawke*, was lost to a submarine attack early
on. To take the blockading ships further away from the submarine threat,
their patrol lines were moved out of the North Sea, which fortuitously
worked to their advantage. The blockading ships 'at once [...] observed
that the shipping was far more numerous than they had met further south'.
However, the sea is much rougher to the north, which was to the 10CS's
detriment. Rapidly it became obvious that these elderly cruisers were far
from the ideal vessels for a blockade in northern waters. The major problem
was their mechanical unreliability due to their age, but their poor seaworthiness came a close second. *Crescent* had the indignity of having not only
her admiral's sea cabin washed over the side in rough weather (fortunately
the admiral was on the bridge at the time), but also her boats, which were
essential for boarding blockade runners. By November the decision was
taken to replace the cruisers with armed merchant cruisers. Some of the
original ships went on to serve at Gallipoli, but *Gibraltar* remained with

the squadron, albeit reduced to being a depot and guard ship at a coaling base established for the squadron at Swarbucks Minn in the Shetlands.[19]

The armed merchant cruisers that were used to reconstitute 10CS were requisitioned liners commissioned under the white ensign. They were available because the war had significantly reduced oceanic passenger traffic, and they had many advantages over the cruisers. They were much more seaworthy, and many were faster than the warships they replaced. They were newer and had much more reliable machinery; above all they had a greater coal capacity, meaning that they could stay far longer at sea. Despite this the *Viknor* and *Clan MacNaughton* were lost very early on due to bad weather. They did have the disadvantage, as Admiral de Chairs's correspondence with the Admiralty repeatedly mentioned, that they were not warships and thus very susceptible to battle damage and consequent flooding.

The ships were armed with a mixture of 6-inch and 4.7-inch guns. The actual number of ships varied, but by January 1915 the 10CS was made up of 23 ships. All were commanded by Royal Naval officers, and their crews were part naval (in some cases men from the *Edgars*) and part civilian under naval command serving under merchant navy contracts known as 'T124' articles. Curiously, despite being warships, the ships' owners were responsible for provisioning them.

The Northern Patrol, as it became known, operated by patrolling along a designated part of one of a series of patrol lines until they either encountered a ship by chance or, as was more often the case, were directed to a blockade runner by wireless. The patrol lines varied in detail throughout the war, but Fig. 7.2 shows the patrol lines as they were in January 1915.

The blockading ships were kept informed of merchant ship movements by the Admiralty; Room 40 and the war room tracked merchant ships as well as warships. The information was surprisingly detailed even from the start of the war and became more so as the Admiralty's intelligence machinery matured. When a ship of the patrol sighted a suspect ship, it would be ordered to stop. It would then be boarded and inspected, and allowed to proceed if it was not found to be carrying contraband. If there was any suspicion that the ship was carrying contraband, it would be ordered into port, usually Kirkwall, for further inspection and prize adjudication. In most cases this would be with a party from the patrol on board. If the cargo was deemed contraband by the prize court it was confiscated.

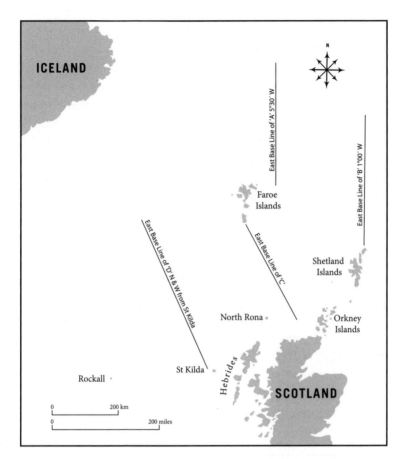

Fig. 7.2. The patrol lines as they were in January 1915. They did not
vary much throughout the existence of the Northern Patrol.

It was not only cargo that the inspecting officers were looking for.
Particularly in the early stages of the war there were many military reservists
attempting to return to Germany. Military personnel and reservists were
liable to arrest and internment under international agreement. To try to
circumvent internment, the German consulate in New York established
an office to issue false travel documents: most commonly the travellers
purported to be Norwegian.[20]

What did not help matters was that the rules to which 10CS had to
work were being constantly changed. As noted earlier, on 20 August 1914
a British government Order in Council had said that the Declaration of

London would be adhered to, subject to some modifications. One modification was the introduction of the 'doctrine of continuous voyage', notwithstanding the provisions of Chapter I, Article 19. In essence, contraband was contraband if its ultimate destination was Germany, even if it was to transit another country to get there. The reason for this is well illustrated by the soaring imports of cotton by neutral countries neighbouring Germany in the first nine months of the war (see Table 7.1). Cotton was essential for the manufacture of gun cotton, an explosive. It was a major US export, and America was just coming out of a recession that had been particularly bad in the cotton-producing southern states. US cotton farmers would potentially suffer if they could not export to Germany.

Table 7.1. American cotton exports to neutral countries trading with Germany[21]

Exported to	Bales of cotton exported during the 12 months prior to August 1914	Bales of cotton exported between August 1914 and April 1915
The Netherlands	105,000	419,370
Sweden	100,000	747,630
Norway	13,689	125,510
Denmark	27,500	65,830

It is obvious that at the outbreak of war either those neutral nations had suddenly developed a new major usage for cotton, or else they were re-exporting to Germany. Britain was careful not to suggest the latter openly, and initially allowed the imports to those nations to continue to pass through the blockade to avoid upsetting the US government. Other ways of circumventing the blockade were to unload the cargo partially in a neutral port and then to proceed to a German port. There were instances in the Baltic of prearranged 'capture' by German warships. As the United States became more aligned with the Allies, eventually declaring war on Germany, the restrictions imposed by the blockade became tighter.

The British prize court itself was careful to be seen as independent of the government. Casting back to the judgements of their predecessors during the Napoleonic Wars, they stated very firmly that '[t]he idea that the King in Council, or indeed, any branch of the Executive, has power to prescribe or alter the law to be administered by Courts of Law in this country is out of harmony with the principles of our Constitution'. Ultimately this obvious independence and acknowledgement of international law worked to Britain's advantage when the Germans and their submarine campaign flouted agreements they had freely entered into before the war. An early instance was of a German submarine sinking SS *Trondhjemsfjord*, a Norwegian America Line ship, in July 1915 because she had (legally) transferred from the British to the Norwegian flag during the war. This attracted unfavourable Norwegian comment. German submarines also attacked and sank ships that were being taken under escort into Kirkwall – that is to say, those that were carrying possible contraband to Germany.[22]

Once the prize court had been set up and consequent rules and regulations implemented so that the Northern Patrol complied with international law, its work in many respects became routine; however, as the area of sea where the Atlantic Ocean and the North Sea come together is some of the roughest in the world, it was very wearing on the crews and ships. Storm damage, and even loss, was commonplace.

The ships had to contend not only with the 'dangers of the sea' but also the 'violence of the enemy', as the naval prayer has it. It was known in February 1916 that a German surface raider, the SMS *Möwe*, would attempt to return to Germany from a raiding trip in the Atlantic. HMS *Alcantara,* a former Royal Mail Steam Packet, was on patrol and had been warned by the Admiralty that a surface raider might be encountered. Another ship, the HMS *Andes*, came on what appeared to be a Norwegian merchant ship, the *Rena*, but her particulars did not appear in *Lloyd's Confidential List of Ships*. She was in fact the SMS *Greif*, an outbound German raider. *Alcantara* came up in support and opened fire on the *Greif*. In the ensuing action, both *Alcantara* and *Greif* were sunk, *Alcantara* losing five officers and 115 men.

Initially the War Committee, an inner cabinet, oversaw the blockade, but its importance was recognized in February 1916 when a new Ministry of Blockade was set up under Robert Cecil, who was given a seat in the Cabinet. Admiral de Chair handed over direct responsibility for the

Fig. 7.3. The action between the HMS *Alcantara* and the SMS
Greif. The *Greif*, as is portrayed, disguised herself as a Norwegian
merchant ship, but fought under the German naval ensign.

Northern Patrol to Vice Admiral Tupper and became the naval adviser to
the new ministry. Cecil set about 'rationing the imports of neutral countries in contact with Germany'. At the same time he arranged that Britain
would purchase their exports. He was able to threaten them subtly with
the control of essential imports such as coal and sugar, 'always on condition
that [they didn't] send it on to Germany'.[23]

The patrol was increasingly effective. In the second quarter of 1916
only 34 out of 916 ships going either to or from Germany's neutral neigh-
bours evaded the patrol. The blockade was tightened by the use of more
ships supported by trawlers and by the use of Halifax, Nova Scotia, as an
examining port. Restrictions on coal supplies to neutral vessels made for
better control of their movements. Gradually a system of 'navicerts' was
introduced. These were issued by British consuls overseas and effectively

allowed neutrals to 'clear customs' before departure. Once the United States entered the war in April 1917, the seaborne element of the blockade was largely redundant; neutral shipping was controlled from the ports. By June 1917 the ships of 10CS were being transferred to convoy duties.[24]

Britain's blockade of Germany worked and, as will be described, was far more effective than Germany's attempted blockade of Britain. At the end of the first period of unrestricted submarine warfare, the British blockade had stopped 743 neutral ships carrying supplies to Germany, three times as many ships as the Germans sank in the same period. That the blockade was effective is evidenced in 1916 by one woman attempting to carry two tons of rubber in her designated personal baggage in an attempt to evade search. Even by 1916, basic textiles were in short supply in Germany and there was a shortage of fats for foodstuffs. By the end of the war individual Germans were going hungry and industry was very short of raw materials.[25]

Not only did the blockade reduce imported food and coal supplies to Germany, but domestic production of food also fell, largely because of the induction of farmers and horses into military service. The combination of this and a poor potato harvest led to the 'turnip winter' of 1916–17. As the war progressed the civilian population was increasingly affected by food shortages. There was increasing malnutrition and related disease. Germany was besieged and being starved. Inevitably this impacted on industry as well as the morale and fighting capability of the army. Such were the shortages facing the German army that during the major German attacks in the spring of 1918, parts of the *Michael* and *Georgette* offensives were slowed because the German troops 'who had suffered so many years of shortages' stopped to loot food and the 'abundance of British supplies did nothing to maintain [German] morale'.[26]

The economic war was far from one-sided. From the outset the German navy attempted to destroy Britain's seaborne supplies of food and materials for her industries, the lifeline that enabled her to fight the war.

The German blockade of Britain

Typically, it was Admiral Fisher who as early as 1904 said: 'I don't think it is even faintly realized – the immense impending revolution which the submarines will effect as offensive weapons of war.'

Germany had actually planned to use surface raiders for their war on trade; some were actual warships, some converted merchant ships. Indeed some of their merchant ships, such as the *Cap Trafalgar*, carried guns in packing cases ready to be mounted so that they could become raiders if war broke out. While they did cause the Allies some problems, the surface raiders actually did not substantially damage trade; it was the submarine that was to do that.[27]

It was obvious from early on that developing submarine technology made them the better weapon for the war on trade that Germany envisaged. Thus its war on British seaborne trade was largely conducted by submarine at varying levels of intensity and at various times during the war, culminating in unrestricted submarine warfare in 1917.[28]

The Germans came relatively late to submarines. *U-1*, their first experimental *Unterseeboot* ('undersea boat', or U-boat), went to sea in 1907 using a kerosene engine with all the attendant risks of fire. It was not until 1913 that their first diesel-powered ocean-going submarine was to appear. Their armament was originally torpedoes; only later did submarines mount a deck gun for use when surfaced. The Germans entered the war with 29 submarines and their Austrian allies had nine; the Royal Navy had 72. Germany's problem was, as ever, geographical. In order actually to get at the mass of merchant shipping in the Atlantic they had to pass through the English Channel or go north around the British Isles.

Before the war, while the British government and the Admiralty were well aware of the submarine's potential, they did not see that it could be used against trade. They felt that the constraints of international law rendered the submarine ineffective as a weapon against merchant shipping. In particular Article 50 of the London Declaration of 1909 required that before any warship destroyed a merchant vessel 'all persons on board must be placed in safety'. It was felt that a submarine would have to make its presence known to achieve this, and in so doing it would lose its major advantage of invisibility. Thus when *U-17* sank the SS *Glitra* on 20 October 1914, it did so by stopping and then boarding her after allowing her crew to escape; only then did it sink her. The whole process took two hours, during which time the submarine was stopped on the surface. Similarly, before being sunk by HMS *Highflyer* on 26 August, the auxiliary cruiser SMS *Kaiser Wilhelm der Grosse*, a surface raider in the Atlantic, had been forced to let two ships go free because it was unable to meet the requirement for the safety of their civilian crew and passengers. However, the Germans were soon to ignore the niceties of the London Declaration.

On 26 October 1914, *U-24* attacked the *Amiral Ganteaume*. The U-boat captain later claimed that he had taken it to be an escorted troopship – a legitimate target if true. The *Amiral Ganteaume* did not sink, but 40 civilian passengers, Belgian refugees, were killed. There followed within Germany a surprisingly public debate about submarine warfare, with public opinion and members of the Reichstag supporting a war on merchant shipping. '[T]he emergence of an unrestricted submarine campaign against shipping was inevitable, and the only thing holding it back was the legacy of nineteenth-century warfare as defined by nineteenth-century technology.'[29]

On 4 February 1915 Admiral von Pohl, commander in chief of the *Hochseeflotte*, announced that after 17 February the waters around the United Kingdom (which then included the whole of Ireland) were to be considered a war zone and any merchant ships, including neutral vessels, were liable to be sunk without warning. The first period of unrestricted submarine warfare had started. While there were those on the Allied side, such as Admiral Fisher, who felt that '[t]he essence of war is violence. Moderation in war is imbecility', Britain was able to use German actions as effective propaganda because they and the United States felt that international efforts to moderate the excesses of war (such as the Hague Conventions) still mattered. It was to be international pressure that led to the suspension of the first period of unrestricted submarine warfare in September 1915.[30]

The Royal Navy was now faced with the problem of defending not only warships but also merchant ships against submarine attack. Some measures were quite simple. Because a submarine's speed and endurance when submerged were actually quite limited, the appearance of a warship would force a submarine to submerge. Once underwater, while invisible, it was less able to attack because its speed and endurance were very much reduced. The merchant ships could repeatedly alter their course, zigzagging at unpredictable intervals, which would complicate the submarine's already difficult job of aiming its torpedoes or, in the modern jargon, achieving a fire control solution. While fast, with speeds of up to 40 knots, the range of a torpedo was limited and they ran in a straight line. Because they used an internal combustion engine for propulsion, they left a visible wake on the surface.

Initially the British tried using warships to patrol along merchant shipping routes. This proved to be ineffective; the submarine would just wait, submerged, until the patrol had passed by. What was lacking was a means of both detecting and sinking a submerged submarine. Submarines on the surface could be sunk by gunfire and by ramming. The latter was even effective against a submerged submarine at periscope depth, which for German submarines meant that the upper part of the pressure hull was only about 10 or 12 feet below the surface, with the conning tower less than that. *Dreadnought* drew 29 feet, quite sufficient on 15 March 1915 to sink *U-29*, which was commanded by the same Otto Weddigen who had sunk *Hawke*, *Hogue*, *Aboukir* and *Cressy*. While ramming was used as an anti-submarine technique well into World War II, it did have the disadvantage of being likely to damage the ramming ship; it was only a partial answer to the problem.

Moored mines were seen as a possible solution. The British were confident that once fully in place, the combination of minefields and the surface Dover Patrol would effectively block the English Channel; in reality, however, it was little more than an inconvenience, and U-boats continued to transit the English Channel until the end of the war, losing six submarines during 253 passages from January to November 1917.[31]

To sink a submarine, the ideal method was to hit it with an explosive charge. The Royal Navy had been experimenting with towed sweeps since 1910. The idea was simple: a hydroplane with an explosive charge was towed by a ship in such a way that it 'flew' underwater, out from the side of the ship. If it hit a submarine, it would explode. While it was a good idea and widely used, it was largely ineffective. Some submarines were hit by them, but none were sunk. Various other means of deploying an explosive charge, such as through howitzer-type guns, were tried before the depth charge evolved. The depth charge was essentially a canister full of explosives with a fusing device that set it off at a preset depth. It could be deployed simply by dropping it into the water, either from the stern of a ship or later from an aircraft, or even by propelling it a distance away from a ship by a depth-charge thrower, so that a spread of charges could be dropped on a submarine. Because a single depth charge had a limited effective radius, a pattern was more likely to be effective if the position of the submarine was not accurately known, much as a poor shot is more likely to hit a target with a shotgun than with a rifle.

There remained the problem of finding a submerged submarine and localizing it well enough to attack it. Detection loops were used at major bases such as Scapa Flow. Laid on the sea bottom, these were cables that relied on the inherent magnetism of a submarine affecting an electric current flowing in the cables. These were often combined with controlled mines, so the operator ashore could explode a mine close to the presumed submarine. However, these were complex and were only of use around major harbours and bases. Similarly, nets were a useful defence, but mainly for static targets such as a port or anchored warships. Suspended from floats, they could form an effective barrier.

What was needed was a method of detecting submarines that would be a threat to a ship at sea rather than ones in harbour. Detector nets, analogous to a fisherman's float, were use with limited success in the Straits of Dover, but the real advances were made using hydrophones, literally listening for submarines. Sound travels much better, albeit slower, through water than it does through air, and the sounds made by the diesel engines of a surfaced submarine or the electric motors of a submerged submarine could be heard over significant distances. Initially in 1915 the Royal Navy introduced a non-directional hydrophone (it could hear a submarine, but gave no indication of range or bearing), which was lowered over the side of a ship. The ship had to stop, or else the water rushing past would deafen it. By 1917 directional hydrophones were in service, but they suffered from the disadvantage that they could not differentiate whether a sound was coming from one direction or its exact opposite. Two ships working together would obviate this error. To get over the need to stop or slow down, a variety of towed hydrophones were introduced. As well as being employed by ships, hydrophones were used by aircraft. A flying boat would alight on the water and use a hydrophone to locate a submarine. By the end of the war, non-rigid airships were using hydrophones while in flight. This idea was not utilized again until the advent of the anti-submarine helicopter some 50 years later.[32]

Fig. 7.4. The crew of a *Felixstowe* flying boat using a hydrophone.

Work on active detection systems using a quartz crystal to emit a 'ping' and listening for the echo (later to be known as ASDIC and then sonar) started during the war but did not come to fruition until after it. Until they came into being, the navy tried the centuries-old technique of deception and false colours: in this case, warships pretending to be merchant ships. As soon as the Germans started using U-boats to attack merchant ships, some merchant ships were armed with guns. So that this was not seen to be an offensive weapon, the gun was usually placed on the stern of the ship. This was taken a stage further with the use of 'Q' ships. Whether they were so-called because they were originally based in Queenstown (now Cobh) in Ireland or because the original ships were given 'Q' pennant numbers is now uncertain. They appeared to be merchant vessels, but had naval crews and were fitted with concealed weapons, torpedoes and guns. The 'Q' ship would attempt to lure a submarine into surfacing and then reveal itself as a warship. The trick still worked even when the Germans were aware of their existence; because of the restricted numbers of torpedoes the U-boats could carry, they would prefer to sink a merchant vessel using their deck gun. 'Q' ships, however, could only have a limited effect on the German submarine campaign. As with the British blockade, it was to be the reaction of the neutral United States that was to have the greatest effect.

Fig. 7.5. A *Flower*-class sloop. This is HMS *Arabis*, used as an anti-
submarine sloop. Some of her sister ships were used as 'Q' ships.

Unrestricted submarine warfare

The unrestricted submarine campaign caused significant diplomatic upset.
The loss of *Lusitania*, which was sunk on 7 May 1915 by *U-20* along with
many American passengers, had caused a diplomatic upset between the
United States and Germany that had not been resolved when the *U-24*
sank the *Arabis*, 'precipitat[ing] another crisis in the diplomatic relations
between Germany and the United States'.

There was an internal debate in Germany. On the one hand the naval
staff increasingly saw submarines as being the weapon that would win the
war; on the other the Chancellor, Bethmann-Hollweg, held out against
unrestricted submarine warfare, largely because it would upset neutral
countries, especially the United States, whose ships and people were
victims. The Germans now gave U-boat construction priority over new
capital ships, suspending construction of ships already being built and even
of one that had been launched.

While the debate continued, the Germans temporarily moved the focus
of their attack to the Mediterranean, partly because there were likely to
be fewer US citizens, and at the same time gave an undertaking to avoid
attacking passenger ships. This was honoured in the breach rather than
the observance when *U-28* sank the P&O liner *Persia* on 30 December.

After an attack on the Channel packet *Sussex* on 24 March 1916, killing many American citizens, the United States government threatened to break off diplomatic relations with Germany, and so the campaign was suspended on 20 April. Despite heavy losses, by the end of 1916 the British merchant marine was still 94 per cent of the size it had been at the beginning of the war. The reality was that the Germans were not sinking enough ships to impose real economic pressure. However, they still felt that they had a potentially war-winning weapon, yet they were also aware that unrestricted submarine warfare would antagonize the United States. After major debates in Berlin, Germany reinstituted unrestricted submarine warfare on 1 February 1917. The rate of sinkings immediately rose dramatically.[33]

Table 7.2. Allied merchant tonnage losses during 1917[34]

Month	Allied merchant tonnage losses
January	328,391
February	520,412
March	564,497
April	860,334
May	616,316
June	696,725

While the Germans lost only nine submarines in the period, annual wastage of Allied ocean-going tonnage was nearly 23 per cent. Such losses were unsustainable, even after the Allies were reinforced when the United States declared war on 6 April 1917. This was finally precipitated by the sinkings and their associated deaths of US citizens, and by the disclosure of the infamous 'Zimmerman telegram' decrypted by Room 40, in which Germany offered parts of Texas to Mexico if they would declare war on the United States.[35]

The entry of the United States to the war obviously made blockading Germany much easier, and would provide immense resources of manpower and material for the Western Front, but it offered little immediate help to the U-boat problem. Looking back, the solution was obvious: sailing merchant ships in escorted convoys. However, this was not so obvious at the time, nor was convoy easy to implement for a host of reasons.

A convoy was defined by the Admiralty as being 'one or more merchant ships or auxiliaries sailing under the protection of one or more warships'. The navy had historically always relied on convoy to protect trade. During the Napoleonic Wars it was actually a legal requirement for British merchant ships to travel in convoy. During the nineteenth century and the early years of the twentieth, trade protection was much discussed and assumed greater importance as the the likely enemy France, judging from her warship-building programme, intended to conduct any future war primarily against trade. In 1902 the Admiralty formed a 'Trade Division', specifically to look at the defence of maritime trade. It was felt that as it was likely to be surface warships attacking shipping, convoying merchant ships was adjudged as being unlikely to be a helpful system of defence, as a cruiser could easily sink multiple targets.[36]

Before the war broke out most people, if not all, including Churchill, were agreed that convoy was outdated and unnecessary. One fact often forgotten is that while unescorted ships were not supposed to be attacked by submarines, the 1907 Hague Convention allowed merchant ships in convoy to be attacked without warning by submarines. It was also felt that modern communications precluded secrecy of assembly and sailing of a convoy. Another factor was that ports probably could not deal quickly with the simultaneous arrival of a large number of ships, rather than the steady flow of peacetime. Churchill hoped that 'this cumbrous and inconvenient measure [would] not be required'.[37]

The first period of unrestricted submarine warfare had gone reasonably well from the Allies' point of view, and Germany had abandoned it, reverting to a vague compliance with international agreements. When the second period of unrestricted warfare started, Germany had more and better submarines.

In 1917 anti-submarine warfare was in its infancy, which meant that convoy escorts would have very limited ways to sink an attacking submarine. Thus the idea of convoy smacked of putting a lot of eggs in a poorly protected basket. In short, at that time the risks of convoy were thought to outweigh the benefits by far, because the mathematical tools of operational analysis had yet to be developed. Another reason for resisting the full-scale introduction of convoy was insufficient escorts. Troopships were always convoyed, because they were legitimate targets in or out of convoy.

However, as sinkings increased to the point where the German block-ade of Britain was beginning to work, the defence of trade had to be re-examined. Early on the British listening service realized that U-boats radioed a daily report to their commanders. This meant that close inshore, particularly in the Irish Sea or the North Sea, it was possible to obtain a relatively accurate fix on her. This allowed the possibility of routing convoys and ships away from submarines. This was the forerunner of the techniques used by the Admiralty in World War II.[38]

This exposed a problem. The Admiralty did not wish to disclose the basis for its routing instructions, and as a result merchant captains were inclined to discount such advice. In 1915 the master of the *Lusitania* had 'disobeyed an Admiralty instruction to zigzag [issued] because a U-boat was nearby. (Her captain wanted to reach port on time.)' He, his ship and many passengers paid the ultimate price. Evasive routing did reduce losses, but the German reaction was to move their U-boats closer inshore and thus closer to the nodes, or in modern parlance 'choke points', that ships had to pass through.[39]

Convoy moved up the agenda and was discussed in detail at an Allied naval conference in September 1917. Before operational analysis tools were available, it was not seen that convoy would make large tracts of the ocean empty of ships, and if one submarine did happen on a convoy, the escorts that could easily sink a surfaced submarine would force it to attack while submerged. This would give it limited time and weaponry to sink ships. Indeed the American Admiral Sims actually put forward the notion that convoy was in fact an offensive measure. However, even he sounded a note of caution that convoys would be a tempting target for surface ships – precisely the reason that the Admiralty had turned against them before submarines became a threat.

Convoy was introduced piecemeal. Admiral Dare at Milford had on his own initiative introduced convoys in his command area as early as December 1916.[40] In March 1917 coal-trade convoys were introduced for the cross-Channel coal trade. These were called 'controlled sailings' rather than convoy; this was a legalism to avoid the use of the term convoy, which would permit German attacks on them. Likewise, Scandinavian trade was convoyed from April 1917.[41] The French followed suit for their coastal traffic, and in May 1917 a 16-ship escorted convoy sailed from Gibraltar to England without loss. That same month North Atlantic outbound traffic was convoyed; UK-bound convoys started from August.

Unfortunately the introduction of convoy has attracted a near mythology centred on David Lloyd George, the Prime Minister. According to (his) legend, he arrived in the Admiralty on 30 April 1917, appropriated the First Lord's chair in the Admiralty boardroom and laid down the law. If this version is to be believed, as a result, convoy was introduced over the objections of the Admiralty Board. As with many of Lloyd George's memories, it is not supported by the recollections of others, even those of Maurice Hankey, his cheerleader. As has been discussed previously, the convoy system had already been introduced anyway. However, this did not enhance Admiral Jellicoe's standing in the eyes of Lloyd George, particularly as it came immediately after his failed machinations to weaken and marginalize Field Marshal Haig at the Calais conference in the days before.[42]

Convoys required escorts, and there were simply not enough ships. Initially the Royal Navy used destroyers as being the best type of ship available, but this was at the expense of the Grand Fleet. Destroyers were an integral part of the Grand Fleet, not only as escorts for the battleships and battlecruisers, but to strike at the enemy's capital ships. As a first step to deal with the shortage of escorts, fishing trawlers were commandeered. Between 1914 and 1945 over 1,000 trawlers and smaller drifters were either commandeered from the fishing fleet or specially built. They were armed, but their size meant that they were often actually less well armed than a surfaced German submarine, and they could carry very little in the way of anti-submarine weapons such as depth charges. The British also built a series of 1,200-ton *Flower*-class sloops. Some were completed to look like merchant ships and employed as 'Q' ships. They were built to mercantile rather than warship standards and were built extremely quickly; the first was ordered in December 1914 and was commissioned, having completed her sea trials, on 22 May 1915.[43]

As a more definitive escort, the Royal Navy also ordered a class of 'P' (for patrol) boats. Like World War II *Flower*-class corvettes, these 613-ton vessels were designed to be built cheaply and rapidly out of mild steel (except for the bow, which was made of hardened steel for ramming) by

Fig. 7.6. *P59*, one of the mass-produced anti-submarine patrol boats.

shipbuilders who did not normally undertake warship work. Ultimately, 64 were built between 1915 and the end of the war. They were of simple lines and had a very low silhouette, which was actually similar to that of a surfaced submarine, and were propelled by two steam turbines. Originally they carried only two depth charges, but as the realities of anti-submarine warfare became apparent, this was increased to 30. Use of these and other classes freed many destroyers for their primary intended purpose, but destroyers were still used because there were never enough escorts.[44]

The major effect of the introduction of convoy was, from the U-boats' perspective, to empty the ocean. The smoke from a single ship could be seen for up to 20 miles, but the smoke from a convoy not from much further. There would now be one column of smoke from a convoy of 40 ships rather than 40 separate columns spread out over the ocean.

Once the Germans had declared unrestricted submarine warfare in British waters, 'the first twelve months of war had made the south Irish waters of the Atlantic a veritable graveyard for shipping of all sorts – from the world's biggest steamers down to sailing craft and submarines.'[45] The Admiralty decided to appoint a single commander to oversee the area which otherwise was the responsibility of various local flag officers. Admiral Bayly had before the war been an officer who was well thought of; he had preceded Beatty in command of the 1st Battlecruiser Squadron. In July 1915 he was asked by Arthur Balfour, First Lord of the Admiralty, to be 'Senior Officer on the Coast of Ireland', a position that became eventually 'Commander in Chief, Coast of Ireland', with responsibility for the whole of the coast of Ireland, the Scillies, the Irish Sea and even Milford Haven in South Wales. In fact in July 1917 the title of his post was changed to 'Commander in Chief, Western Approaches'. He was based in Ireland at Queenstown, as until 1922 the whole of the island of Ireland was part of the United Kingdom and there were naval bases at Queenstown in the south and at Berehaven on Bantry Bay in the south-west. From these bases ships had ready access to the south-western approaches to the United Kingdom and the Atlantic.[46]

Bayly was an inspired choice. Despite a reputation for being brusque, even rude, he was very popular, particularly after increasing numbers of United States navy destroyers – ultimately 56 – were placed under his command following America's entry into the war in 1917. He built up the command, initially using *Flower*-class sloops and Royal Navy destroyers, as well as US destroyers. It is worth repeating part of the signal that Bayly made in September 1917 to all of his ships, as it gives a flavour of the man and what his command had undertaken:

> The Commander-in-Chief wishes to congratulate Commanding Officers on the ability, quickness of decision, and willingness they have shown in their duties of attacking submarines and protecting trade.[47]

As well as surface vessels, submarines were deployed from Berehaven in an anti-submarine role. Originally this was undertaken by British submarines (under the command of Captain Martin Nasmith, VC) and then after 6 March 1918 by seven US navy submarines which maintained their patrols until almost the end of the war. It was reported that the Germans found the presence of opposing submarines as well as surface vessels unnerving. Nonetheless, after the war the commanding officer of a U-boat asked to join the association of Queenstown commanding officers formed by the US and Royal navies on the basis that he too had served in the Queenstown command area!

Queenstown command and convoys demonstrated that it was not necessary to sink submarines to break the German blockade. It was the safe and timely arrival of the convoy that was a victory; sinking a submarine was a bonus. The combination of central control of shipping, what would later be termed 'evasive routing', the use of convoy, steadily improving sensors and weapons, including aircraft, and new tactics eventually first broke the morale of the German submarine force and then broke the German blockade. Before that stage could be reached, the Germans reacted by moving their submarines closer inshore to the choke points where there were likely to be concentrations of ships. By the end of the war, U-boats were operating in groups in deeper waters, with a command ship controlling them by wireless, an early progenitor of World War II wolf-packs.

New ways to defeat the U-boats

The northern route into the Atlantic, while far longer, was thought to be safer for German submarines than the English Channel. In 1917 the British suggested and the United States navy largely implemented the 'Great Northern Barrage', a minefield across the North Sea from Orkney to the Norwegian coast at Utsira, close to Bergen. It sank up to eight U-boats, and possibly a further eight were sunk or damaged.

Another technology completely new to maritime warfare in World War I was that of aircraft. Lighter than air machines, non-rigid, semi-rigid and rigid airships had made their appearance in a naval guise before the outbreak of war and were used for scouting. Their potential for anti-submarine warfare was appreciated very early on, and they were used as part of the protection for the British Expeditionary Force as it was transported to France in 1914. Initially their limitations, particularly mechanical reliability and weather conditions, restricted their use. This meant that aeroplanes proper were only used for coastal patrols initially; crews were unsurprisingly nervous to be too far from land with an engine that might stop unexpectedly. With improving reliability came greater and wider use.

Ultimately, aircraft were seen by U-boat commanders as a major threat. Not only were they able to attack submarines with depth charges, but their mere presence, even if they did not see the submarine, would force a submarine to dive and thus lose any speed advantage they had over a ship or convoy when surfaced. Even when submerged a submarine could be seen from the air if it put its periscope up, causing a wake, or 'feather', and in calm water it might even be visible itself.

Fig. 7.7. The non-rigid *SSZ 37* overflying a merchantman. This was the best class of airship in World War I.

Fig. 7.8. An aerial photograph of a World War II vintage submerged submarine with its periscope up through the surface. The wake, or 'feather', can be seen, and in the calm sea so can the submarine.

If a submarine fired a torpedo when submerged, its easily visible track
in the water would literally point an aircraft to the submarine. By the end
of the war it was felt that the ideal escort for a convoy should include 'at
least two aircraft, one keeping close and one cruising wide to prevent a
submarine from getting into a position to attack'.[48]

Once fully implemented, convoy was very successful at protecting
against submarine attack. As the American Admiral Sims had predicted,
the Germans now used surface ships to attack convoys. The worst such
episode from the British point of view occurred on 17 October 1917.
The Germans used two fast (34-knot) light cruisers, SMS *Bremse* and SMS
Brummer, to attack a coal convoy from England to Norway. The convoy of
12 ships was escorted by two destroyers and two armed trawlers. Both
escorting destroyers, HMS *Mary Rose* and HMS *Strongbow*, together with
nine of the merchantmen, were sunk without the Germans sustaining any
losses. Fortunately the Germans did not follow up this success; the ever-
present Grand Fleet was a major deterrent to the use of surface ships in
the North Sea. Reaching the Western Approaches with surface warships
would have been impossible.[49]

Who won the trade war?

The Germans lost two campaigns at sea. Almost as important as the trade
war itself was what would now be called the public-relations battle, which
boiled down to 'poor, put-upon Germany insisting on its rights' versus
'the beastly Hun murdering innocent civilians'. Ultimately Germany lost
this, the all-important battle for American public opinion, and brought
the United States into the war.

World War I itself was won at sea, but it was not won by the battlefleets;
it was won by the Northern Patrol under Admiral de Chair and his successor
Admiral Tupper, who took over in 1916, and in the Western Approaches
by the Queenstown Command. In particular Admiral Sir Lewis Bayly,
who commanded in Queenstown, is now almost completely forgotten,
but he ultimately did far more to defeat Germany than many much more
famous admirals and laid the foundations for future cooperation between
the Royal and United States navies.

CHAPTER 8

The Royal Navy in World War I

I have peace to weigh your worth now all is over…

Sub-Lieutenant Rupert Brooke, RNVR
Anson Battalion, Royal Naval Division

In March 1918 the Germans risked everything on an offensive on the Western Front against the British sector, primarily on the Fifth Army. A sortie by the German fleet at about the same time came to nothing, and the British army held, just, and returned to the offensive in August. The Germans fell back and their retreat rapidly became a rout. The German army was soon on their border. The end of the war in November 1918 came far sooner than most on the Allied side had contemplated. Planning was already well under way for the 1919 offensives on the Western, Italian and Balkan fronts. The Grand Fleet had already embarked torpedo-carrying Sopwith *Cuckoo*s in HMS *Argus* for a carrier strike against the German fleet in Wilhelmshaven. However, the German armed forces collapsed; a final sally by the *Hochseeflotte* was abandoned when the ships' companies mutinied. Thus faced with mutinies in its army and its navy, the Germans asked for an armistice, which came into effect at 11:00 on 11 November 1918.

As part of the armistice, the *Hochseeflotte* was to be interned at Scapa Flow, the Grand Fleet's anchorage in the Orkneys. They were received theatrically by the Grand Fleet commanded by Admiral Beatty. Once they were at anchor, he issued an order, not actually covered by the armistice agreement, but which was obeyed nonetheless: 'The German flag will be hauled down at sunset today […] and will not be hoisted again without permission.'[1]

The final act of the naval war came on 21 June 1919 when the Germans scuttled their fleet. The reasons behind their doing this are complex and

Fig. 8.1. The *Hochseeflotte* being escorted into Scapa Flow
to be interned, having steamed the greatest distance it
had ventured from its German bases in wartime.

Fig. 8.2. SMS *Derfflinger* sinking on 21 June 1919,
having been scuttled by her crew.

closely linked to the peace negotiations at Versailles. It meant there was
no possibility of the naval war being resumed if the treaty talks had broken
down and the armistice ended. For the Allies, the naval war was won.

The Royal Navy shared the relief and exultation of the nation that the
war was over and had been won. However, there was a feeling of dissatisfac-
tion, a feeling that the navy had not played its part, that the army had won
the war and that, at least to some extent, the navy had been bystanders.

> The mood of the Royal Navy and the Admiralty as the war neared its end was
> grim. There was 'a feeling of incompleteness', to quote the First Sea Lord.
> The Navy had, in a sense, won a victory greater than Trafalgar, but it was less
> spectacular – there had been no decisive sea battle.[2]

Such feelings went beyond the service, and the public opinion was that:

> [d]espite a good deal of patriotic bluster, the war had done the image of the
> navy no good. The vastly expensive fleet of dreadnoughts had not deterred war.
> It had failed to bring the enemy to a decisive action, as the navy had allowed
> the public and the press to expect.[3]

In a way the problem was the education of the public, and indeed of the
navy itself. As the Royal Navy had expanded at the end of the nineteenth
century, it did so in parallel with the popularity of a series of books by an
American naval theoretician, Captain Alfred Thayer Mahan. The first of
these was *The Influence of Sea Power on History*. It is difficult over a century
later to appreciate the effect his books had; they ran into many editions,
and even Kaiser Wilhelm had studied them. Mahan's view was believed to
be that the battlefleet was the key factor in sea power and that ultimately
naval wars were decided by fleet actions. His later book *The Influence of
Sea Power on the French Revolution and Empire, 1793–1812* emphasized this
point. The Royal Navy had expected a 'Glorious First of June', a Nile and
a Trafalgar, and felt that it had failed because it had not annihilated the
German fleet at Jutland.[4]

Before World War I Julian Corbett, who would go on to be the author
of the official history of World War I at sea, was preaching a very different message, but his ideas had not received much attention. His 'Green

Pamphlet' of January 1909, described as being 'Strategical Terms and Definitions used in Lectures on Naval History', emanated from the Royal Naval War College at Portsmouth as part of a book that is still cited today and is even currently 'enjoying a renaissance among strategists of the [Chinese] People's Liberation Army Navy'. Corbett preached that seapower was about economic power, not just about battlefleets. When the war is examined as it is hoped this book has done, Corbett was right; however, the wider navy, and indeed the country, did not immediately fully appreciate that it had played a very major part in bringing Germany to its knees. Furthermore, despite initial political vacillation, it had done this largely in the manner it had intended before the war: by economic means. It is the nature of war plans that they are kept secret; the Admiralty did not announce, even to its own people, that it planned to strangle Germany rather than beat her to death, and so in large part its case has been ignored, particularly by politicians. This was very much a case where perception trumped reality, and was fertile ground for internal critics of naval education such as Richmond.[5]

The use of sea power

After the formation of the Royal Air Force in 1918, it became a mantra for the new service that 'air power is indivisible'. This was and is patent nonsense. Air defence of the United Kingdom is obviously different from strategic bombing, which is different again from air defence of a convoy. The only thing they have in common is the use of aircraft. However, seapower *is* indivisible, and World War I perfectly illustrates this fact. Unfortunately the war at sea tends historically to be divided up for ease of description; for example, the Grand Fleet tends to be looked at separately from the war on trade, without the linkage that the former enabled the latter. This missing link hides the interrelation between the different elements of seapower used to defeat the enemy.

The blockade played a large part in the defeat of Germany. However, it did so using armed merchant cruisers, liners with naval crews and few guns. The loss of the *Alcantara* when fighting the *Greif* showed how vulnerable they were, and experience in World War II with the losses of HMS *Jervis Bay* and HMS *Rawalpindi* to German surface warships[6] showed just how easily sunk they were. Both Admiral de Chair and his successor Admiral Tupper recognized that a single major German warship could by itself have broken the blockade by sinking the entire 10th Cruiser Squadron in a very short time. There was, however, one very big 'if'. In order to get to the blockade lines, such a ship would have had to get past the Grand Fleet: 'the armed merchant cruisers did the actual work, but it was the menacing presence of those great dreadnoughts, always ready to intervene if challenged, which enabled them to work undisturbed.'[7]

It was economic pressure that finally broke Germany. By the end her people were starving, and they, both civilian and military, were mutinous. 'What ifs' are unproductive; history is what happened, and no amount of arguing can change it. War in all its aspects is unpleasant, but if the Admiralty had been allowed to prosecute the war as it wished, and as the Cabinet had approved in 1908, Germany would have starved sooner. Ultimately, probably far fewer lives, service and civilian, on both sides would have been lost, but it would have incurred the enmity of neutral nations, particularly the United States, which would have had their trade disrupted.

The ships

The German ships, both battlecruisers and battleships, have been described as better armoured and possessed of better guns and particularly shells. In fact there is no doubt at all that British ships carried larger guns and had a far greater weight of broadside (the weight of all the shells fired from the guns simultaneously). In later classes of ships the British superiority in weight of broadside was marked. *Queen Elizabeth* fired a broadside of 15,600 lbs, which was 7,000 lbs more than her German contemporary the *Kronprinz Wilhelm*, which had a nominally identical armament of eight 15-inch guns.

Particular opprobrium has been heaped on the British battlecruisers because they were too lightly armoured. The question of poor turret and magazine practices has been dealt with already. There is a strong possibility that this at the least contributed to the loss of three battlecruisers. Certainly *Lion* at the battle of Dogger Bank had her armour penetrated, as did *Tiger* at Jutland, but it did not sink them. It must be remembered that no armour could make a ship invulnerable anyway. Nor is this one-sided:

at both battles German ships had their armour penetrated, and Fig. 8.3 shows the effects of British gunfire on the *Seydlitz*.

It is curious that the debate has continued for so long. After the German fleet was interned and scuttled at Scapa Flow the opportunity arose to examine the German ships. One ship examined in great detail was *Baden*, generally held to be the ultimate in German battleship design; indeed her design was the basis for that of the *Bismarck* of World War II. She was almost exactly contemporary with the HMS *Revenge* of the 'R' class. The R-class battleships were an economy version of the *Queen Elizabeth*-class that preceded them. *Baden* and *Revenge* were almost identical in size, armament and machinery. However, *Baden*'s construction was lighter, and the German naval architects appear to have accepted stresses up to 20 or 25 per cent more than was allowed by the Royal Corps of Naval Constructors; put simply, she was more flexible – not an ideal characteristic for a warship. This design philosophy probably accounts for the *Seydlitz*'s leaking due to shock damage to her hull simply caused by firing her main armament. In terms of stability, *Baden* was slower to roll, having a higher metacentre – ideal as a gun platform in the North Sea – but she would have been at a disadvantage in the longer swells of the North Atlantic. While her subdivision

Fig. 8.3. SMS *Seydlitz* in harbour after the battle of Jutland.

was good in terms of her having four longitudinal bulkheads running the length of the ship, the transverse bulkheads running from side to side had numerous piercings for pipe work and ventilation, conferring a greater risk to her watertight integrity. Add to that her magazines and shell rooms, which, while behind armour, extended far closer to the ship's side than did *Revenge*'s, making them more vulnerable. *Baden*'s guns could only elevate to 16°, giving her a maximum range of 22,200 yards. For *Revenge* it was 20° and 25,000 yards; her guns could engage a target nearly two land miles further away. By any measure, *Revenge* was the better ship.[8]

In terms of gunnery, the Germans were widely held to have better rangefinders, using the stereoscopic principle rather than the British coincidence principle. However, in trials after the war the British rangefinders proved superior. British fire control using director firing – all the guns being controlled from a single point high up in the ship – was far superior to the German system of a centralized measurement of range being passed to the gun turrets, which then fired independently.

There was, however, one element of gunnery practice where the Germans were superior. Inevitably the range measured by optical rangefinders was an approximation. Once measured, the guns needed to 'find the range' exactly. Where the Germans were better was in their technique of 'ladder' shooting. They would fire a rapid sequence of salvoes, increasing or decreasing the gun range in stages, which is analogous to climbing or descending a ladder. This meant that they found the range quicker than the British system, in which of a series of ranging salvoes would be fired and the range then adjusted after each shot fell. This was slower than firing a ladder. Not surprisingly, the British switched to 'ladder' shooting.

The big British weakness was their shells. If they hit obliquely, their armour-piercing shells would fail to do just that. This was discovered after Jutland, so new and improved shells were rapidly issued. Apart from that, in terms of material the Royal Navy and its ships were much better than the Germans', and after a century this should now be recognized.

The people

The Royal Navy was fortunate that it was able to fight for much of the war with long-service personnel. Unlike the army, which expanded enormously

Fig. 8.4. Admiral Sir John Jellicoe.

and also suffered appalling casualties, the navy did not have the great training burden, except in the Royal Naval Division. Quite rightly no one has questioned the quality, training, professionalism and dedication of the lower deck of the navy. They remained that way throughout the war despite being paid badly, particularly in comparison with better-paid industrial workers, who lived at home and were spared the long boring periods at anchor or on repeated sweeps, looking for the Germans who seemed never to come to sea. The British sailor in World War I was every bit as good as his predecessors and his successors in World War II.

The junior officers as a group have been criticized, but for factors beyond their remedy. Churchill and Admiral Richmond were contemporary critics, and many historians such as Correlli Barnett have followed to such an extent that it is now largely accepted that officers at all levels were hidebound, rigid, unthinking, poorly educated automatons. If that is accepted then the supposed 'failings' of the Royal Navy in World War I can be laid at their door. This is a complete travesty. Firstly, the navy did not fail. It must not be forgotten that 20 years later many of the selfsame officers were to be responsible for fighting World War II, and the navy certainly did not fail in that war, as even Correlli Barnett accepts. Leopards do not change their spots.[9]

It is useful to break down the officers' performance, taking leadership first. It has already been said that the ratings performed extremely well and could only have done so if they were well led. Thus their officers were good leaders. They showed plenty of courage; think of Harvey, Loftus-Jones and Nasmith, to name but three Victoria Cross winners. It is a long list, and it extends far beyond winners of the Victoria Cross. Turning to initiative: this has already been discussed in Chapter 5, but the point bears repeating that officers had both to impose and to practice rigid and at times unthinking obedience, and they had to be ready when required to show immediate initiative. There can be no better example than the officers of the destroyers at Jutland.

However, a lot of criticism can be justly levelled at some of the senior officers, captains and admirals, whose performance varied from the truly outstanding to the dismal and even stupid.

At one end of the scale is Jellicoe, who has come in for a lot of ill-informed and prejudiced criticism. He was held by Fisher, no mean judge of a naval officer, to be the best of his generation. He was extremely popular at every level, and popular because he was seen as being a good and extremely able leader. He undoubtedly had his faults; many authors have remarked on his unwillingness to delegate. However, as he showed at Jutland, he would defer to expert opinion. He was quite rightly brought to London as First Sea Lord in 1916. While two years of command at sea would have taken its toll, as First Sea Lord he proved to be dynamic and capable, undertaking necessary and overdue reorganizations of the Admiralty. Much opprobrium has been heaped on him for his 'failure' to implement convoy. As has been shown, he could not have done so any earlier than he did, because there were not enough escorts. Convoy was implemented sooner than those who have sought to decry him have admitted, particularly Lloyd George. Lloyd George makes much of having sacked Jellicoe for this failure. Even the Welsh wizard's cheerleaders such as Hankey do not agree with his account. The reality is probably that Lloyd George, like many politicians before and since, was taken with dramatic gestures. He had after all founded a new fighting service for no real reason other than to deflect blame from himself; he also tried to sack Field Marshal Haig and failed. Sacking Jellicoe may well have been a substitute.

It has been suggested that the Royal Navy was overburdened with poor-quality senior officers who had been promoted because there were no alternative candidates when the navy expanded rapidly in the 20 years leading up to the war. That does not explain the rise of Admiral Beatty. Beatty was promoted by the system and through the system, because the system picked him out for rapid promotion. He was a brave and inspiring leader of men, and his service on the Nile during the Sudan Campaign got him a very early promotion to commander and a DSO. He was then appointed to *Barfleur* as the commander (executive officer, or second in command); she was a flagship, and this would have been Beatty's only experience of seeing a flag and staff at work at sea until he went to sea as an admiral. It was cut short by his taking part in another predominantly land-based operation; this time in China with the Boxer Rebellion. He did well, so well in fact that with no more than two years in rank he was promoted to captain at the age of 29. Command of three ships followed. Throughout this period he performed exceptionally well, and on 1 January 1910 became the youngest flag officer since the eighteenth century.

Fig. 8.5. Admiral Sir David Beatty.

He then went on the senior officers' war course. At the end of the course he was one of the most junior admirals in the navy, but asked for commands that were appropriate to someone more senior. When he was offered the appointment as second in command of the Atlantic fleet, he declined it. He was told in no uncertain terms that this was unacceptable, and there his career might have ended had he not at Churchill's request become his naval secretary. During this period he was given command for six weeks of a squadron of old armoured cruisers for the annual naval exercise. He flew his flag in *Aboukir*, which was to be his only sea experience as an admiral until in 1913, at Churchill's insistence, he was given command of the battlecruisers.[10]

Beatty was undoubtedly an officer of the highest ability; he was an able ship commander; the ships that he commanded performed very well. However, when he was given command of the battlecruisers he was woefully inexperienced in flag command; he had done an admiral's job at sea for only six weeks before taking command of the most important scouting group in the Grand Fleet. Normally an admiral would at least expect to have served in a flagship for a full commission of two years and seen, even if as a ship's officer and an outsider, how the staff worked. Best of all he would have served as a flag captain under an admiral and then would have been a subordinate flag officer at sea in a post such as the one he refused. His performance as a flag officer bore out his inexperience. At the battle of Dogger Bank he divided his command team when he returned to the compass platform in *Lion* with his secretary and flag lieutenant, leaving the remainder in the conning tower. At the same battle his subordinates certainly did not think the way he wanted, despite his being in command for nearly two years. His tactical handling of the 5th Battle Squadron at Jutland was woeful, but his greatest weakness was in keeping Lieutenant Commander Seymour as his flag lieutenant. The flag lieutenant was Beatty's mouthpiece to his subordinate commanders, and Seymour was not up to the job. It will be remembered that after the Scarborough raid, Beatty tried to have Admiral Goodenough sacked for what in reality was Seymour's failing; Beatty should have sacked Seymour for that, then again for his signalling at Dogger Bank, and yet again at Jutland. Beatty commented later that Seymour had lost him three battles; perhaps he would have been a better admiral if he had sacked Seymour after the first.

Jellicoe and Beatty were far from being the only flag officers of the Royal Navy during World War I whose performance warrants examination. It must not be forgotten that the Admiralty had a hierarchy of important appointments and put its best officers in them accordingly. Those perceived as the brightest and the best were put in the most important posts. Tyrwhitt, Goodenough and Keyes filled absolutely pivotal appointments as commodores, although they were still substantive captains at the outbreak of war. Goodenough retired as a full admiral (four star, in modern parlance); the other two made admiral of the fleet (five star). Admiral Bayly had commanded a battle squadron before being chosen to go Queenstown, which proved to be an outstanding choice.

Admiral Cradock has attracted little but faint disparagement (more than that from Churchill) for being beaten. In fairness he did his best with what he had in order to carry out his duty, which is more than can be said for Troubridge, who appears to have failed by thinking too much. Arbuthnot, however, thought too little. Commanding an armoured cruiser squadron should not have taxed his intellect too far, but he successfully got two ships sunk unnecessarily.

Probably the three most important admirals for their contribution to the Allied victory were Admirals de Chair and Tupper, who commanded the sea blockade of Germany, and Bayly, who was pivotal in defeating the German submarine threat. If Jellicoe and Beatty are dimly remembered, these three are now largely forgotten, and if those admirals are forgotten, the Admiralty and the naval staff are only remembered to be criticized. It cannot be emphasized too much that the Admiralty in the first decade of the twentieth century established the first operational naval staff in history and learnt very quickly how to use it. It brought together a range of new technologies and techniques to fight a naval war. This was the first organization ever to integrate what in modern military parlance is known as C3I – Command, Control, Communications and Intelligence – and their use at every level of command. It made mistakes, much as would any new organization using new techniques, and was heavily dependant on its people. Historians have made much of the failure by the Admiralty to tell Jellicoe that the Germans were at sea before Jutland. This was due to a mistake by one officer about the German use of callsigns. But this did not matter. The system coped with the mistake, and it should not be

forgotten that because of intelligence the Grand Fleet was at sea before
the Germans (to deal with a German sortie), and use of local communications intelligence meant that Jellicoe knew that the *Hochseeflotte* was
at sea in good time.

One person who played a major part in the management of the Royal
Navy in the years immediately before World War I and into 1915 warrants
particular attention: Winston Churchill. If, as seemed likely at the time,
his political career had ended in 1915, he would now be remembered as a
brash, interfering politician, who in overreaching himself, failed. In looking
at his performance from the perspective of the twenty-first century, the
retrospectoscope (an easily and much-used historical instrument) inevitably
also takes in the years of his incomparable greatness in World War II. That
should not be allowed to obscure his earlier years, which may be thought
of as his time as an apprentice warlord. Churchill was extremely intelligent and possessed an incisive mind. It must not be forgotten that when
he came to the Admiralty as First Lord, it was as an excited young politician who was climbing the greasy pole and had been given an important
Cabinet appointment. He spent a lot of time listening to Fisher, which can
be thought of as preparing himself, but with his military experience, albeit
at a junior level, if he had been given the War Office, he would have been
seen to be equally prepared, indeed even more so. At his age Churchill
had the arrogance of a young, highly intelligent man with no respect for
the older men in the navy, and set about wholesale sackings on his arrival.

Churchill was publicly determined to establish a naval staff. The records
show that after a year he tacitly admitted that there was a very effective
naval staff already in existence when he arrived. He took over the reins
of the Royal Navy and had the grace to admit later that he was making
operational decisions that were properly the responsibility of the First Sea
Lord. Indeed there are signals in the records annotated in his hand that
they were to be shown to the First Sea Lord *after* dispatch. When Fisher
resigned, effectively taking Churchill with him, Jellicoe wrote to Fisher:

> We owe you a debt of gratitude for having saved the Navy from a continuance
> in office of Mr Churchill, and I hope that never again will any politician be
> allowed to usurp the functions that he took upon himself to exercise.

Churchill made some bad mistakes, and spent a lot of time after the war justifying his actions. A lot of his book *The World Crisis, 1911–1918* is just that. It is still today an extremely readable book, and it must not be forgotten that it was written by what is today almost unknown: a financially honest and broke politician, who needed to sell the book and was bound to be contentious in order to do so. He was an apprentice warlord, and like all apprentices he made mistakes, some very bad ones. Unfortunately he did take forward some misapprehensions; in particular he underestimated the effects of the war on trade, both on Britain and on Germany, and he certainly was at first to underestimate the U-boat threat in World War II.[11]

Conclusion

Britain had historically avoided land involvement on the continent of Europe as far as was possible. Before World War I, the biggest threat that Germany was seen to pose to Britain was at sea, and the navy was seen as its most important defence. The Admiralty had planned on waging a largely economic war on Germany. The battlefleet was essential because the Germans had one, and it had to be matched if Britain was to be sure of command of the sea so that it could wage economic warfare. In 1911 the change to that policy certainly came as a surprise to the First Sea Lord, Admiral Wilson, and it cost him his job.

Once war came, the navy was not seen to be 'doing enough'. The dull, dreary business of blockade did not attract public attention, certainly not while the British army was bleeding in France. Perhaps the politicians ought to have listened to the words of the Royal Navy's quick march 'Heart of Oak', sung on patriotic occasions, which include: 'Why if they won't fight us, we cannot do more.' Germany, and more particularly the Kaiser, believed in a 'fleet in being'; he was not prepared to risk his pet toy, and there was no final battle, no maritime Armageddon. The result was the continuing perception that the Royal Navy contributed little to the winning of World War I, whereas it was largely responsible for Germany's defeat, and that defeat was due to British seapower.

Notes

List of bibliographical abbreviations:

AMO = Admiralty Monthly Order; AWO = Admiralty Weekly Order; BR = Book of Reference; CB = Confidential Book; KR & AI = Kings Regulations and Admiralty Instructions

Introduction

1. Thucydides, quoted in Victor Davis Hanson, *A War Like No Other* (New York: Random House, 2005), p. 10.

Chapter 1. Europe on the Brink of War

1. Robert K. Massie, *Castles of Steel: Britain, Germany and the Winning of the Great War at Sea* (London: Jonathan Cape, 2004), p. 130.
2. Nicholas A. Lambert, *Planning Armageddon: British Economic Warfare and the First World War* (Cambridge, MA: Harvard University Press, 2012), p. 170.
3. Annika Mombauer, *Helmuth von Moltke and the Origins of the First World War* (Cambridge: Cambridge University Press, 2001), p. 287.
4. Arthur J. Marder, *From the Dreadnought to Scapa Flow. Volume 1: The Road to War, 1904–1914* (Barnsley: Seaforth Publishing, 2013), pp. 143–4.
5. Correlli Barnett, *The Swordbearers: Studies in Supreme Command in the First World War* (London: Eyre and Spottiswoode, 1963), p. 185.

Chapter 2. The Royal Navy at the Outbreak of World War I

1. Andrew Gordon, *The Rules of the Game: Jutland and British Naval Command* (London: John Murray, 1996), p. 155.
2. Oscar Parkes, *British Battleships:* Warrior *1860 to* Vanguard *1950. A History of Design, Construction and Armament* (London: Leo Cooper, 1990), p. 531.
3. Brian Lavery, *Able Seamen: The Lower Deck of the Royal Navy, 1850–1939* (London: Conway, n.d.), p. 234.
4. Christopher M. Bell, 'The King's English and the security of the empire: class, social mobility, and democratization in the British Naval Officer Corps, 1918–1939', *Journal of British Studies* 48 (July 2009), pp. 695–716; KR & AI, Appendix XII, part 3; KR & AI, article 261.

5. AWO 2195/18, *Acting Mate – Promotion to*; Navy List (London: HMSO, December 1918). The DSC was awarded for 'gallantry in the face of the enemy'.

6. W.S. Galpin, *From Public School to Navy: An Account of the Special Entry Scheme* (Plymouth: Underhill, 1919), p. 10; see also John H. Beattie, *The Churchill Scheme: The Royal Naval Special Entry Cadet Scheme, 1913–1955* (private publication, 2010), which covers the same ground in much greater detail.

7. H.W. Dickinson, *Educating the Royal Navy: Eighteenth- and Nineteenth-Century Education for Officers* (London: Routledge, 2007), p. 205; M.A. Farquharson-Roberts, 'To the Nadir and back: the executive branch of the Royal Navy, 1918–1939', unpublished Ph.D. thesis (University of Exeter, 2013).

8. Christopher Page, *Command in the Royal Naval Division: A Military Biography of Brigadier General A.M. Asquith, DSO* (Staplehurst: Spellmount, 1999).

9. S.A. Cavell, *Midshipmen and Quarterdeck Boys in the British Navy, 1771–1831* (Woodbridge: The Boydell Press, 2012), p. 153.

10. Richard Freeman, *The Great Edwardian Naval Feud: Beresford's Vendetta against Fisher* (Barnsley: Pen and Sword Maritime, 2009).

11. BR 1875 (previously CB 3013), *Naval Staff Monograph (Historical): The Naval Staff of the Admiralty, Its Work and Development* (London: Admiralty, 1929), pp. 24–9; C.I. Hamilton, *The Making of the Modern Admiralty: British Naval Policy-Making, 1895–1927* (Cambridge: Cambridge University Press, 2011), pp. 219, 243.

12. Barry D. Hunt, *Sailor-Scholar: Admiral Sir Herbert Richmond, 1871–1946* (Waterloo, Ontario: Wilfred Laurier University Press, 1982); N.A.M. Rodger, 'Training or education: a naval dilemma over three centuries', in P. Hore (ed.), *Hudson Papers 1* (London: Oxford University Hudson Trusts, 2001), pp. 23–4.

13. Farquharson-Roberts, 'To the Nadir and back'; Nicholas Black, *The British Naval Staff in the First World War* (Woodbridge: Boydell and Brewer, 2009), p. 14.

14. Dwayne R. Winseck and Robert M. Pike, *Communication and Empire: Media Markets and Globalization* (Durham, NC: Duke University Press, 2007), pp. 24–5; Norman Friedman, *Network-Centric Warfare: How Navies Learned to Fight Smarter through Three World Wars* (Annapolis, MD: Naval Institute Press, 2009), p. 3.

15. Friedman, *Network-Centric Warfare*, p. 7.

16. CB 1681A, *Grand Fleet Battle Orders* (London: Naval Staff, 1919).

17. Hugh Cleland Hoy, *40 O.B., or How the War Was Won* (London: Hutchinson, 1932); Friedman, *Network-Centric Warfare*, pp. 3–13.

18. Friedman, *Network-Centric Warfare*, pp. 6–7.

19. E.H.H. Archibald, *The Fighting Ship in the Royal Navy, AD 897–1984* (New York: Military Press, 1987), p. 196.

20. Lord Hankey, *The Supreme Command, 1914–1918* (London: George Allen and Unwin, 1961), p. 118.

21. Winston S. Churchill, *The World Crisis, 1911–1918* (London: Penguin Classics, 2007), p. 99.

22. Ibid., p. 58; TNA/PRO ADM 116/3060.

23. Churchill, *The World Crisis*, p. 120.

24. William Guy Carr, *Brass Hats and Bell-Bottomed Trousers: The Unforgettable and Splendid Feats of the Harwich Patrol* (London: Hutchinson, 1939).

25. Arthur J. Marder, *Portrait of an Admiral: The Life and Papers of Sir Herbert Richmond* (London: Jonathan Cape, 1952), p. 293.

26. Paul G. Halpern, *A Naval History of World War I* (London: UCL Press, 1994), p. 35.

27. Philip Ziegler, *Mountbatten: The Official Biography* (London: Collins, 1985), p. 36.

28. Robert K. Massie, *Castles of Steel: Britain, Germany and the Winning of the Great War at Sea* (London: Jonathan Cape, 2004), p. 165; Richard Hough, *The Great War at Sea, 1914–1918* (Oxford: Oxford University Press, 1983), pp. 100–2; Halpern, *A Naval History*, p. 35; Churchill, *The World Crisis*, p. 219.

29. Churchill, *The World Crisis*, p. 131.

Chapter 3. Towards Armageddon

1. Christopher Clark, *The Sleepwalkers: How Europe Went to War in 1914* (London: Penguin, 2013), p. 469.

2. Oscar Parkes, *British Battleships: Warrior 1860 to Vanguard 1950. A History of Design, Construction and Armament* (London: Leo Cooper, 1990), p. 603; Winston S. Churchill, *The World Crisis, 1911–18* (London: Penguin, 2007), pp. 106–7.

3. Julian S. Corbett, *Naval Operations: History of the Great War Based on Official Documents* (Uckfield: Naval and Military Press, n.d.), vol. 1, pp. 60–1.

4. Norman Friedman, *Network-Centric Warfare: How Navies Learned to Fight Smarter through Three World Wars* (Annapolis, MD: Naval Institute Press, 2009), p. 7.

5. P.K. Kemp, *The Papers of Admiral Sir John Fisher* (n.l.: Navy Records Society, 1960), pp. 161–2.

6. A. Berkeley Milne, *The Flight of the 'Goeben' and the 'Breslau': An Episode in Naval History* (London: Eveleigh Nash, 1921), p. 120.

7. E.H.H. Archibald, *The Fighting Ship in the Royal Navy, AD 897–1984* (New York: Military Press, 1987), pp. 201, 204.

8. Robert K. Massie, *Castles of Steel: Britain, Germany and the Winning of the Great War at Sea* (London: Jonathan Cape, 2003) pp 26–55; Richard Hough, *The Great War at Sea 1914–1918* (Oxford: Oxford University Press, 1983), pp. 69–87; Paul G. Halpern, *A Naval History of World War I* (London: UCL Press, 1994), pp. 51–64; Archibald, *The Fighting Ship*, p. 238.

9. William Guy Carr, *Good Hunting* (London: Hutchinson, 1940), p. 9.

10. *Landrail*'s gun mounting is in the National Museum of the Royal Navy in Portsmouth.

11. Halpern, *A Naval History*, p. 27.

12. Roger Keyes, *The Naval Memoirs of Admiral of the Fleet Sir Roger Keyes: The Narrow Seas to the Dardanelles, 1910–1915* (London: Thornton Butterworth, 1934), p. 81.

13. Halpern, *A Naval History*, p. 31.

14. Ibid., pp. 30–2; Massie, *Castles of Steel*, pp. 96–121; Carr, *Brass Hats*, pp. 44–71; Hough, *The Great War*, pp. 65–8; Archibald, *The Fighting Ship*, pp. 195, 207–8.

15. Ian Buxton, *Big Gun Monitors: The History of the Design, Construction and Operation of the Royal Navy's Monitors* (Tyne & Wear: World Ship Society / Trident Books, 1978), pp. 87–8.

16. Geoffrey Bennett, *Coronel and the Falklands* (London: Pan Books, 1967); Hough, *The Great War*, pp. 87–98; Halpern, *A Naval History*, p. 76; Massie, *Castles of Steel*, pp. 179–97; Churchill, *The World Crisis*, pp. 226–50.

17. Corbett, *Naval Operations*, vol. 1, p. 442.

18. Bennett, *Coronel and the Falklands*, pp. 124–6.

19. Massie, *Castles of Steel*, pp. 280–1.

20. Corbett, *Naval Operations*, vol. 2, pp. 249–51.

21. Barbara W. Tuchman, *The Guns of August* (New York: Ballantine Books, 2004), pp. 372–82.

22. Corbett, *Naval Operations*, vol. 2, p. 28.

23. Ibid., pp. 28, 38–9; Massie, *Castles of Steel*, p. 348.

24. Corbett, *Naval Operations*, vol. 2, pp. 82–102; Massie, *Castles of Steel*, pp. 375–425; Halpern, *A Naval History*, pp. 44–7; Hough, *The Great War*, pp. 130–43.

25. Friedman, *Naval Weapons of World War One: Guns, Torpedoes, Mines and ASW Weapons of All Nations* (Barnsley: Seaforth, 2011), pp. 23–6.

26. Filson Young, *With the Battle Cruisers* (Memphis, TN: General Books, 2012), pp. 48, 51.

27. Andrew Gordon, *The Rules of the Game: Jutland and British Naval Command* (London: John Murray, 2000), p. 94.

28. Admiral James Goldrick, personal communication to the author, 6 September 2013.

29. Massie, *Castles of Steel*, p. 416.

Chapter 4. The Amphibious Navy

1. See Britt Zerbe, *A History of the Royal Navy: The Royal Marines* (London: I.B. Tauris, 2014).

2. See Andrew Baines, *A History of the Royal Navy: The Victorian Age* (London: I.B. Tauris, 2014).

3. Douglas Jerrold, *The Royal Naval Division* (Uckfield: Naval and Military Press, n.d.), p. 2.

4. See Chapter 6.

5. Robert K. Massie, *Castles of Steel: Britain, Germany and the Winning of the Great War at Sea* (London: Jonathan Cape, 2003), pp. 164–5.

6. Geoffrey Sparrow and J.N. Macbean Ross, *On Four Fronts with the Royal Naval Division during the First World War, 1914–1918* (n.l.: Leonaur, 2011), p. 21.

7. Winston S. Churchill, 'Introduction', in Douglas Jerrold, *The Royal Naval Division* (Uckfield: Naval and Military Press, n.d.), p. xvii; Christopher Page, *Command in the Royal Naval Division: A Military Biography of Brigadier General A.M. Asquith, DSO* (Staplehurst: Spellmount, 1999), p. 30.

8. Jerrold, *The Royal Naval Division*, p. 44.

9. Ibid., p. 48.

10. Arthur J. Marder, *From the Dreadnought to Scapa Flow. Volume 1: The Road to War, 1904–1914* (Barnsley: Seaforth Publishing, 2013), p. 148.

11. Paul G. Halpern, *A Naval History of World War I* (London: UCL Press, 1994), p. 110.

12. Julian S. Corbett, *Naval Operations: History of the Great War Based on Official Documents* (Uckfield: Naval and Military Press/Imperial War Museum, n.d.), vol. 2, pp. 110–18.

13. Halpern, *A Naval History*, p. 110.

14. Page, *Command in the Royal Naval Division*, p. 42.

15. Michael Hickey, 'Gallipoli: the Constantinople expeditionary force, April 1915', in T.T.A. Lovering (ed.), *Amphibious Assault: Manoeuvre from the Sea: Amphibious Operations from the Last Century* (London: MoD, 2005), pp. 7–21.

16. R.H. Gibson and Maurice Prendergast, *The German Submarine War, 1914–1918* (London: Constable, 1931), p. 71; Corbett, *Naval Operations*, vol. 2, p. 203, n. 1.

17. Halpern, *A Naval History*, p. 119.

18. Richard Harding, 'The long shadow of the Dardanelles on amphibious warfare in the Royal Navy', in Peter Hore (ed.), *Dreadnought to Daring: 100 Years of Comment, Controversy and Debate in the Naval Review* (Barnsley: Seaforth, 2012), pp. 83–94.

19. Page, *Command in the Royal Naval Division*, p. 73.

20. Ibid., pp. 89–90.

21. Gary Sheffield, *Forgotten Victory: The First World War: Myths and Realities* (London: Review, 2002), p. 197; Robert L. Davison, *The Challenges of Command: The Royal Navy's Executive Branch Officers, 1880–1919* (Farnham: Ashgate, 2011), p. 92.

22. Jerrold, *The Royal Naval Division*, pp. 294–5.

23. J.P. Harris and Niall Barr, *Amiens to the Armistice: The BEF in the Hundred Days Campaign, 8 August–11 November 1918* (London: Brassey's, 1998), p. 103.

24. Gibson and Prendergast, *The German Submarine War*, p. 16.

25. HMS *M33* is preserved in Portsmouth Historic Dockyard, adjacent to the National Museum of the Royal Navy.

26. Ian Buxton, *Big Gun Monitors: The History of the Design, Construction of the Royal Navy's Monitors* (Tyne & Wear: World Ship Society/Trident Books, 1978), p. 47.

27. Henry Newbolt, *Naval Operations: History of the Great War Based on Official Documents* (Uckfield: Naval and Military Press/Imperial War Museum, n.d.), pp. 241–65.

28. H.W. Dickinson, 'The Zeebrugge and Ostend raids: Operation ZO, April 1918', in T.T.A. Lovering (ed.), *Amphibious Assault: Manoeuvre from the Sea: Amphibious Operations from the Last Century* (London: MoD, 2005), pp. 37–48; Arthur J. Marder, *Portrait of an Admiral: The Life and Papers of Sir Herbert Richmond* (London: Jonathan Cape, 1952), p. 293.

29. Halpern, *A Naval History*, pp. 411–15.

30. M.A. Farquharson-Roberts, 'To the Nadir and back: the executive branch of the Royal Navy, 1918–1939', unpublished Ph.D. thesis (University of Exeter, 2013), p. 268.

Chapter 5. The Battle of Jutland

1. Robert K. Massie, *Castles of Steel: Britain, Germany, and the Winning of the Great War at Sea* (London: Jonathan Cape, 2004), p. 566.

2. See Chapter 7.

3. Matthew S. Seligmann, *The Royal Navy and the German Threat, 1901–1914: Admiralty Plans to Protect British Trade in a War Against Germany* (Oxford: Oxford University Press, 2012), pp. 64–88; Massie, *Castles of Steel*, p. 568.

4. Andrew Gordon, *The Rules of the Game: Jutland and British Naval Command* (London: John Murray, 1996), pp. 54–6.

5. Eric Grove (ed.), 'The autobiography of Chief Gunner Alexander Grant: HMS *LION* at the Battle of Jutland, 1916', in *Naval Miscellany Volume VII* (Aldershot: Ashgate/Naval Records Society, 2008), p. 389; Norman Friedman, *Naval Weapons of World War One: Guns, Torpedoes, Mines and ASW Weapons of All Nations* (Barnsley: Seaforth, 2011), p. 28.

6. J.E.T. Harper, *The Truth about Jutland* (London: John Murray, 1927), p. 50.

7. Massie, *Castles of Steel*, p. 590.

8. Georg von Hase, *Kiel and Jutland* (London: Skeffington, n.d.), p. 89; Julian S. Corbett, *Naval Operations: History of the Great War Based on Official Documents* (Uckfield: Naval and Military Press/Imperial War Museum, undated), vol. 3, pp. 336–7; Taffrail [Taprell] Dorling, *Endless Story: Being an Account of the Work of the Destroyers, Flotilla-Leaders, Torpedo-Boats and Patrol Boats in the Great War* (London: Hodder and Stoughton, 1932), pp. 156–63.

9. Dorling, *Endless Story*, p. 168.

10. Harper, *The Truth about Jutland*, pp. 69–70.

11. Discussion with Admiral of the Fleet Sir Henry Leach, 28 October 2008.

12. Arthur J. Marder, *From the Dreadnought to Scapa Flow. Volume 3: The Royal Navy in the Fisher Era, 1904–1919: Jutland and After* (London: Oxford University Press, 1966), pp. 101–2, n. 24.

13. Ibid., p. 102.

14. Ibid., p. 98.

15. Ibid., pp. 99–100.

16. Ibid., p. 108.

17. Friedman, *Naval Weapons of World War One*, pp. 337–8.

18. Lord Chatfield, *The Navy and Defence: The Autobiography of Admiral of the Fleet Lord Chatfield* (London: William Heinemann, 1942), p. 148.

19. Massie, *Castles of Steel*, p. 636.

20. Dorling, *Endless Story*, p. 193.

21. Corbett, *Naval Operations*, vol. 3, pp. 400–1.

22. Massie, *Castles of Steel*, p. 646.

23. Gordon, *The Rules of the Game*; Ronald A. Hopwood, *The Old Way and Other Poems* (London: John Murray, 1916), p. 20.

24. Correlli Barnett, *The Swordbearers: Studies in Supreme Command in the First World War* (London: Eyre and Spottiswoode, 1963), pp. 184–5; Winston S. Churchill, *The World Crisis, 1911–1918* (London: Penguin Classics, 2007), p. 620.

25. BR 827, *A Seaman's Pocket-Book* (London: HMSO, June 1943), p. 4; Ian Buxton, *Big Gun Monitors: The History of the Design, Construction and Operation of the Royal Navy's Monitors* (Tynemouth: World Ship Society and Trident Books, 1978), p. 173; Peter Hodges, *The Big Gun: Battleship Main Armament, 1860–1945* (London: Conway Maritime Press, 1981), pp. 133–4; John Campbell, *Naval Weapons of World War Two* (London: Conway Maritime, 1985), pp. 25–8.

26. Seligmann *The Royal Navy and the German Threat*, pp. 64–88.

27. Anthony P. Tully, *The Battle of Surigao Strait* (Birmingham, IN: Indiana University Press, 2009), p. 90.

28. Barrie Kent, *Signal! A History of Signalling in the Royal Navy* (Clanfield: Hyden House, 1993), pp. 61–2.

29. H.W. Dickinson, *Educating the Royal Navy: Eighteenth- and Nineteenth-Century Education for Officers* (London: Routledge, 2007), p. 205.

Chapter 6. The Navy and New Technologies

1. Ronald A. Hopwood, 'The New Navy', in *The New Navy and Other Poems* (London: John Murray, 1919), p. 10.

2. Correlli Barnett, *The Swordbearers: Studies in Supreme Command in the First World War* (London: Eyre and Spottiswoode, 1963), p. 185; Winston S. Churchill, *The World Crisis 1911–1918* (London: Penguin Classics, 2007), p. 620.

3. Bryan Cooper, *The Ironclads of Cambrai* (London: Pan Books, 1970), pp. 12–13; John Glanfield, *The Devil's Chariots: The Birth and Secret Battles of the First Tanks* (Stroud: Sutton, 2001) pp. 57–8.

4. Cooper, *The Ironclads of Cambrai*, pp. 15–16.

5. Christina J.M. Goulter, 'The Royal Naval Air Service: a very modern service', in Sebastian Cox and Peter Gray (eds), *Air Power History: Turning Points from Kitty Hawk to Kosovo* (London: Frank Cass, 2002), pp. 51–65.

6. Niall Barr, 'Command in the transition from mobile to static warfare: August 1914 to March 1915', in Gary Sheffield and Dan Todman (eds), *Command and Control on*

the Western Front: The British Army's Experience, 1914–1918 (Stroud: Spellmount, 2007), pp. 13–38.

7. Terry C. Treadwell, *The First Naval Air War* (Stroud: Tempus Publishing, 2002), pp. 29–31.

8. Julian S. Corbett, *Naval Operations: History of the Great War Based on Official Documents* (Uckfield: Naval and Military Press/Imperial War Museum, n.d.), vol. 2, pp. 51–2.

9. Geoffrey de Havilland, *Sky Fever: The Autobiography of Sir Geoffrey de Havilland, CBE* (n.l.: Wrens Park, 1999), pp. 73–4; A. J. Jackson, *Blackburn Aircraft since 1909* (London: Putnam, 1968), p. 7; J.M. Bruce, *War Planes of the First World War: Fighters* (London: Macdonald, 1965), vol. 1, p. 97.

10. Bruce, *War Planes of the First World War*, vol. 1, pp. 67–76; ibid., vol. 3, p. 67.

11. S.W. Roskill (ed.), *Documents Relating to the Naval Air Service: 1908–1918* (n.l.: Navy Records Society, 1969), pp. 168–9.

12. Bruce, *Warplanes of the First World War*, vol. 2, pp. 131–9.

13. Mark Barber, *Royal Naval Air Service Pilot, 1914–18* (Oxford: Osprey, 2010), pp. 34–6; Bruce, *War Planes of the First World War*, vol. 3, pp. 131–9.

14. Norman Friedman, *Naval Weapons of World War One: Guns, Torpedoes, Mines and ASW Weapons of All Nations* (Barnsley: Seaforth, 2011), p. 328.

15. R.D. Layman, *Naval Aviation in the First World War: Its Impact and Influence* (London: Caxton Editions, 2002), pp. 152–3.

16. John J. Abbatiello, *Anti-Submarine Warfare in World War I: British Naval Aviation and the Defeat of the U-Boats* (London: Routledge, 2011), p. 19.

17. Layman, *Naval Aviation*, p. 80; Abbatiello, *Anti-Submarine Warfare*, pp. 83, 97–8.

18. Abbatiello, *Anti-Submarine Warfare*, pp. 83, 90–2, 97–8.

19. Charles Harvard Gibbs-Smith, *Aviation: An Historical Survey from its Origins to the End of World War II* (London: HMSO, 1970), p. 287.

20. Bruce, *War Planes of the First World War*, p. 87.

21. Oscar Parkes, *British Battleships: Warrior 1860 to Vanguard 1950. A History of Design, Construction and Armament* (London: Leo Cooper, 1990), pp. 622–3; John Roberts, *Battlecruisers* (London: Chatham Publishing, 1997), pp. 53–4.

22. David Hobbs, *Aircraft Carriers of the Royal and Commonwealth Navies* (London: Greenhill Books, 1996), p. 93.

23. Ibid., p. 40.

24. Roskill (ed.), *Documents Relating to the Naval Air Service*, p. 273.

25. J.C. Smuts, *Jan Christian Smuts* (London: Cassell, 1952), p. 205; Layman, *Naval Aviation*, p. 198.

26. Arthur J. Marder, *From the Dreadnought to Scapa Flow. Volume 4: 1917 – Year of Crisis* (London: Oxford University Press, 1967), p. 333.

27. David Lloyd George, *War Memoirs* (London: Odhams, 1936), pp. 1097–1115; W.J. Reader, *Architect of Air Power: The Life of the First Viscount Weir* (London: Collins, 1968), pp. 73–4.

28. Corbett, *Naval Operations*, vol. 3, pp. 288–96.

29. David Hobbs, 'The first Pearl Harbour: the attack by British torpedo planes on the German High Seas Fleet planned for 1918', in John Jordan (ed.), *Warship 2007* (London: Conway, 2007), pp. 29–38.

30. Nicholas Lambert, *The Submarine Service, 1900–1918* (Aldershot: Ashgate/Navy Records Society, 2001), p. x.

31. Peter Hore, 'British submarine policy from *St Vincent* to Arthur Wilson', in Martin

Edmonds (ed.), *100 Years of the Trade* (Lancaster: CDISS, 2001), pp. 11–12; Lambert, *The Submarine Service*, p. xi.

32. See Nicholas Lambert, *Sir John Fisher's Naval Revolution* (Columbia, SC: University of South Carolina Press, 1999), p. 53. The first of the *Holland*-class is displayed in the Royal Naval Submarine Museum in Gosport, a part of the National Museum of the Royal Navy.

33. William Guy Carr, *By Guess and by God: The Story of the British Submarines in the War* (London: Hutchinson, 1930), pp. 61–2.

34. Lambert, *Sir John Fisher's Naval Revolution*, p. 123.

35. Lambert, *The Submarine Service*, p. xix.

36. Jon Davidson and Tom Allibone, *Beneath Southern Seas* (Crawley: University of Western Australia Press, 2005), p. 108.

37. Lambert, *The Submarine Service*, p. xliv; William Guy Carr, *Hell's Angels of the Deep* (London: Hutchinson, n.d.), pp. 208–9; Carr, *By Guess and by God*, p. 141.

38. Carr, *By Guess and by God*, p. 17.

39. Lambert, *The Submarine Service*, p. 265; Henry Newbolt, *Naval Operations: History of the Great War Based on Official Documents* (Uckfield: Naval and Military Press/Imperial War Museum, n.d.), pp. 91–8.

40. Carr, *By Guess and by God*, pp. 234–7.

41. Lambert, *The Submarine Service*, p. xxxii.

42. Edwin Gray, *A Damned Un-English Weapon: The Story of British Submarine Warfare* (London: NEL, 1973), pp. 170–4.

43. Duncan Redford, personal communication to the author, 7 Nov 2013.

Chapter 7. The British and German Wars against Trade

1. Anonymous, 'Special report: world economy – railroads and hegemons', *Economist* (12 October 2013), p. 5.

2. Nicholas A. Lambert, *Planning Armageddon: British Economic Warfare and the First World War* (Cambridge, MA: Harvard University Press, 2012), p. 239.

3. Robert K. Massie, *Castles of Steel: Britain, Germany and the Winning of the Great War at Sea* (London: Jonathan Cape, 2004), pp. 506–7.

4. Lambert, *Planning Armageddon*, p. 41.

5. Andrew Lambert, *The Challenge: America, Britain and the War of 1812* (London: Faber and Faber, 2012), pp. 198–202.

6. 'Declaration concerning the Laws of Naval War, 208 Consol. T.S. 338 (1909)', [online], <http://www1.umn.edu/humanrts/instree/1909b.htm>, accessed 27 April 2013; E. Keble Chatterton, *The Big Blockade* (London: Hutchinson & Co, 1932), p. 31.

7. Stephen Roskill, *Hankey: Man of Secrets, 1877–1918* (London: Collins, 1970), p. 58.

8. Lambert, *Planning Armageddon*, p. 185.

9. Lord Hankey, *The Supreme Command, 1914–1918* (London: George Allen and Unwin, 1961), pp. 160–1.

10. Keble Chatterton, *The Big Blockade*, p. 30.

11. Annika Mombauer, *Helmuth von Moltke and the Origins of the First World War* (Cambridge: Cambridge University Press, 2001), pp. 72–7.

12. John D. Grainger, *The Maritime Blockade of Germany in the Great War: The Northern Patrol, 1914–1918* (Aldershot: Ashgate/Naval Records Society, 2003), p. 8.

13. Lambert, *Planning Armageddon*, p. 355.
14. Hankey, *The Supreme Command*, p. 352.
15. Ibid., p. 353.
16. Admiral Sir Dudley de Chair, *The Sea Is Strong* (London: George Harrap, 1961), p. 160.
17. Alexander Scrimgeour, *Scrimgeour's Small Scribbling Diary, 1914–1916. The Truly Astonishing Wartime Diary and Letters of an Edwardian Gentlemen, Naval Officer, Boy and Son*, eds Richard Hallam and Mark Beynon (London: Conway, 2008), p. 32.
18. Keble Chatterton, *The Big Blockade*, p. 34.
19. Grainger, *The Maritime Blockade of Germany*, p. 58.
20. Ibid., p. 71, n. 1.
21. Keble Chatterton, *The Big Blockade*, p. 117.
22. Ibid., p. 147; Grainger, *The Maritime Blockade of Germany*, p. 245.
23. Hankey, *The Supreme Command*, p. 547.
24. Grainger, *The Maritime Blockade of Germany*, p. 723.
25. Ibid., pp. 338–9; Massie, *Castles of Steel*, p. 545.
26. Gary Sheffield, *Forgotten Victory: The First World War: Myths and Realities* (London: Review, 2002), p. 234.
27. William Guy Carr, *Hell's Angels of the Deep* (London: Hutchinson, n.d.), pp. 125–7.
28. Quoted in Don Everitt, *The K Boats: A Dramatic First Report on the Navy's Most Calamitous Submarines* (London: George G. Harrap, 1963), p. 16.
29. Dwight R. Messimer, *Find and Destroy: Antisubmarine Warfare in World War I* (Annapolis, MD: Naval Institute Press, 2001), p. 17.
30. Lee Willett, 'Submarines in the First World War', in Martin Edmonds (ed.), *100 Years of 'The Trade'* (Lancaster: CDISS, 2001), p. 19.
31. Paul G. Halpern, *A Naval History of World War I* (London: UCL Press, 1994), p. 350.
32. Messimer, *Find and Destroy*, pp. 114–20.
33. Halpern, *A Naval History*, p. 340.
34. Ibid., p. 341.
35. Winston S. Churchill, *The World Crisis, 1911–1918* (London: Penguin Classics, 2007), p. 696; David Kahn, *The Codebreakers* (London: Sphere Books, 1977), pp. 134–53.
36. Roger Knight, *Britain against Napoleon: The Organization of Victory, 1793–1815* (London: Allen Lane, 2013) p. 184; BR 1875 (previously CB 3013), *Naval Staff Monograph (Historical): The Naval Staff of the Admiralty, Its Work and Development* (London: Admiralty, 1929), p. 42.
37. George Franklin, *Britain's Anti-Submarine Capability, 1919–1939* (London: Frank Cass, 2003), p. 113; Christopher M. Bell, *Churchill and Sea Power* (Oxford: Oxford University Press, 2013), pp. 41–2.
38. E. Keble Chatterton, *Danger Zone: The Story of the Queenstown Command* (London: Rich & Cowan, 1934), p. 53.
39. Norman Friedman, *Network-Centric Warfare: How Navies Learned to Fight Smarter through Three World Wars* (Annapolis, MD: Naval Institute Press, 2009), p. 16.
40. John Terraine, *Business in Great Water: The U-Boat Wars, 1916–1945* (London: Leo Cooper, 1989), p. 100.
41. Messimer, *Find and Destroy*, p. 150.
42. David Lloyd George, *War Memoirs* (London: Odhams Press, 1938), p. 692; Roskill, *Hankey*, p. 381; Hankey, *The Supreme Command*, p. 650; Messimer, *Find and Destroy*, p. 152; Walter Reid, *Architect of Victory Douglas Haig* (London: Birlinn, 2006), p. 372.

43. E.H.H. Archibald, *The Fighting Ship in the Royal Navy, AD 897–1984* (New York: Military Press, 1987), p. 245–50.

44. John Roberts, 'P49', *Warship* 12 (October 1979), pp. 252–3.

45. Keble Chatterton, *Danger Zone*, p. 89.

46. Ibid., p. 199.

47. Ibid., p. 298.

48. Terraine, *Business in Great Water*, p. 149; Messimer, *Find and Destroy*, pp. 130–9.

49. Massie, *Castles of Steel*, p. 747.

Chapter 8. The Royal Navy in World War I

1. Arthur J. Marder, *From Dreadnought to Scapa Flow: The Royal Navy in the Fisher Era, 1904–1919* (London: Oxford University Press, 1970), vol. 5, p. 191.

2. Ibid., p. 176.

3. Brian Lavery, *Able Seamen: The Lower Deck of the Royal Navy, 1850–1939* (London: Conway, 2011), p. 265.

4. A.T. Mahan, *The Influence of Sea Power upon History, 1660–1783* (London: Sampson Low, Marston & Co., 1890); Robert K. Massie, *Dreadnought: Britain, Germany and the Coming of the Great War* (London: Pimlico, 1993), p. 256; A.T. Mahan, *The Influence of Sea Power upon the French Revolution and Empire, 1793–1812* (London: Sampson Low, Marston, Searle and Rivington, 1892); Julian S. Corbett, *Some Principles of Maritime Strategy* (London: Filiquarian Publishing, 1911).

5. Corbett, *Some Principles of Maritime* Strategy, p. 143; James R. Holmes and Toshi Yoshihara, 'China's navy: a turn to Corbett', *Proceedings of the United States Naval Institute* 136 (2010), pp. 42–46. Admiral Sir Herbert Richmond (1871–1946) was a naval historian from early in his career and an educationalist. A founder of the *Naval Review*, he established the Senior Officers' War Course at Greenwich and was first commandant of the Imperial Defence College (now the Royal College of Defence Studies) before leaving the service and becoming Vere Harmsworth Professor of Naval History at Cambridge and Master of Downing College. See: Arthur J. Marder, *Portrait of an Admiral: The Life and Papers of Sir Herbert Richmond* (London: Jonathan Cape, 1952); Barry D. Hunt, *Sailor-Scholar: Admiral Sir Herbert Richmond, 1871–1946* (Waterloo: Wilfrid Laurier University Press, 1982).

6. See Duncan Redford, *A History of the Royal Navy: World War II* (London: I.B. Tauris, 2014).

7. Paul G. Halpern, *A Naval History of World War I* (London: UCL Press, 1994), p. 50.

8. Oscar Parkes, *British Battleships: Warrior 1860 to Vanguard 1950. A History of Design, Construction and Armament* (London: Leo Cooper, 1990), pp. 590–1; Norman Friedman, *Naval Weapons of World War One: Guns, Torpedoes, Mines and ASW Weapons of All Nations* (Barnsley: Seaforth, 2011), pp. 43–6.

9. Correlli Barnett, *The Swordbearers: Studies in Supreme Command in the First World War* (London: Eyre and Spottiswoode, 1963), pp. 184–5. See also Correlli Barnett, *Engage the Enemy More Closely: The Royal Navy in the Second World War* (London: Hodder and Stoughton, 1991).

10. Stephen Roskill, *Admiral of the Fleet, the Last Naval Hero: An Intimate Biography* (London: Collins, 1980), pp. 30–58.

11. John Grigg, *Lloyd George: War Leader, 1916–1918* (London: Allen Lane, 2002), pp. 371–2.

Select Bibliography

Unpublished Sources

The National Archives (TNA); Public Record Office (PRO)

Note: In the text of this book these are referred to by file number; thus 'ADM 178/1234', with a title, if given, suffixed.

ADM 1: Admiralty and Ministry of Defence, Navy Department; Correspondence and Papers
ADM 12: Admiralty: Digests and Indexes
ADM 116: Admiralty: Record Office: Cases

Admiralty Historical Branch and Library, Portsmouth

Admiralty Memoranda (AM) (selected) 1916–18; Books of Reference (BR); Confidential Books (CB); Grand Fleet Battle Orders; Fleet Signal Book Volume 1

Unpublished PhD Theses

Davison, Robert L., 'In defence of corporate competence: the Royal Navy Executive Officer Corps, 1880–1919', unpublished PhD thesis (Memorial University of Newfoundland, 2004)

Farquharson-Roberts, Michael, 'To the Nadir and back: the executive branch of the Royal Navy, 1918–1939', unpublished PhD thesis (University of Exeter, 2013)

Jones, Mary, 'The making of the Royal Naval Officer Corps, 1860–1914', unpublished PhD thesis (University of Exeter, 1999)

Rowe, Laura Emily Lucy, 'At the sign of the foul anchor: discipline and morale in the Royal Navy during the First World War', unpublished PhD thesis (Kings College London, 2008)

Yates, James Alexander, 'The Jutland controversy: a case study in intra-service politics with particular reference to the presentation of the Battle-cruiser Fleet's training, conduct and command', unpublished PhD thesis (University of Liverpool, 1998)

Periodicals

Warship (1977–2013)

Books

Abbatiello, John J., *Anti-Submarine Warfare in World War I: British Naval Aviation and the Defeat of the U-Boats* (London: Routledge, 2006)

Archibald, E.H.H., *The Fighting Ship of the Royal Navy, AD 897–1984* (New York: Military Press, 1987)

Bagnasco, Erminio, Enrico Cernuschi and Vincent P. O'Hara, 'Italian fast coastal forces: development, doctrine and campaigns, 1914–1986: from the beginning to 1934', in John Jordan (ed.), *Warship 2008* (London: Conway, 2008), pp. 85–98

Barber, Mark, *Royal Naval Air Service Pilot, 1914–18* (Oxford: Osprey, 2010)

Barnett, Correlli, *The Swordbearers: Studies in Supreme Command in the First World War* (London: Eyre and Spottiswoode, 1963)

—— *Engage the Enemy More Closely: The Royal Navy in the Second World War* (London: Hodder and Stoughton, 1991)

Beattie, John H., *The Churchill Scheme: The Royal Naval Special Entry Cadet Scheme, 1913–1955* (Private Publication, 2010)

Bennett, Geoffrey, *Naval Battles of the First World War* (London: Pan Books, 1976)

Berkeley Milne, Sir A., *The Flight of the 'Goeben' and the 'Breslau': An Episode in Naval History* (London: Eveleigh Nash Company, 1921)

Black, Nicholas, *The British Naval Staff in the First World War* (Woodbridge: Boydell and Brewer, 2009)

Breemer, Jan S., 'The great race: innovation and counter-innovation at sea, 1840–1890', *Corbett Paper No. 2* (The Corbett Centre for Maritime Policy Studies, January 2011)

Brown, David K., *Warrior to Dreadnought: Warship Development, 1860–1905* (London: Chatham Publishing, 1997)

—— *The Grand Fleet: Warship Design and Development, 1906–1922* (Barnsley: Seaforth Publishing, 2010).

Bruce, J.M., *Warplanes of the First World War: Fighters*, in 3 vols. (London: Macdonald, 1965–9)

Buxton, Ian, *Big Gun Monitors: The History of the Design, Construction and Operation of the Royal Navy's Monitors* (Tynemouth: World Ship Society, 1978)

Carr, William Guy, *By Guess and by God: The Story of the British Submarines in the War* (London: Hutchinson, 1930)

—— *Brass Hats and Bell-Bottomed Trousers: Unforgettable and Splendid Feats of the Harwich Patrol* (London: Hutchinson & Co, 1939)

—— *Good Hunting* (London: Hutchinson & Co, 1940)

—— *Hell's Angels of the Deep* (London: Hutchinson, n.d.)

Chalmers, W.S., *Max Horton and the Western Approaches: A Biography of Admiral Sir Max Kennedy Horton GCB, DSO* (London: Hodder and Stoughton, 1954)

Chatterton (see Keble Chatterton)

Churchill, Winston S., *The World Crisis, 1911–1918* (London: Penguin Classics, 2007)

Clark, Christopher, *The Sleepwalkers: How Europe Went to War in 1914* (London: Penguin, 2013)

Corbett, Julian Stafford, *Some Principles of Maritime Strategy* (London: Filiquarian Publishing, 1911)

—— *Naval Operations: History of the Great War Based on Official Documents* (Uckfield: Naval and Military Press, n.d.)

Corrigan, Gordon, *Mud, Blood and Poppycock: Britain and the First World War* (London: Cassell Military Paperbacks, 2003)

Cox, Sebastian and Peter Gray (eds), *Air Power History: Turning Points from Kitty Hawk to Kosovo* (London: Frank Cass, 2002)

Davison, Robert L., *The Challenges of Command: The Royal Navy's Executive Branch Officers, 1880–1919* (Farnham: Ashgate, 2011)

Dewar, Alfred, 'Naval history and the necessity of a catalogue of sources in naval and military essays', in Julian Stafford Corbett (ed.), *Papers Read in the Naval and Military Section at the International Congress of Historical Studies, 1913* (Cambridge: Cambridge University Press, 1914), pp. 83–114

Dickinson, H.W., 'The Zeebrugge and Ostend raids: Operation ZO, April 1918', in T.T.A. Lovering (ed.), *Amphibious Assault: Manoeuvre from the Sea – Amphibious Operations from the Last Century* (Royal Navy: Crown Copyright, 2005), pp. 37–48

Dorling, Taffrail [Taprell], *Endless Story: Being an Account of the Work of the Destroyers, Flotilla-Leaders, Torpedo-Boats and Patrol Boats in the Great War* (London: Hodder and Stoughton, 1936)

Edmonds, Martin (ed.), *100 Years of 'The Trade'* (Lancaster: CDISS, 2001)

Evans, David, *Building the Steam Navy: Dockyards, Technology and the Creation of the Victorian Battlefleet, 1830–1906* (London: Conway Maritime, 2004)

Fisher, Lord, of Kilverstone, *Memories* (London: Hodder and Stoughton, 1919)

Franklin, George, *Britain's Anti-Submarine Capability, 1919–1939* (London: Frank Cass, 2003)

Freeman, Richard, *The Great Edwardian Naval Feud: Beresford's Vendetta against 'Jackie' Fisher* (Barnsley: Pen & Sword Maritime, 2009)

Friedman, Norman, *Network-Centric Warfare: How Navies Learned to Fight Smarter through Three World Wars* (Annapolis, MD: Naval Institute Press, 2009)

—— *Naval Weapons of World War One: Guns, Torpedoes, Mines and ASW Weapons of All Nations* (Barnsley: Seaforth, 2011)

Friedman, Norman (ed.), *Steam, Steel & Shellfire: The Steam Warship, 1815–1905* (London: Conway, 1992)

Galpin, W.S., *From Public School to Navy: An Account of the Special Entry Scheme* (Plymouth: Underhill 1919)

Gibbs-Smith, Charles Harvard, *Aviation: An Historical Survey from its Origins to the End of World War II* (London: HMSO, 1970)

Gibson, R.H. and Maurice Prendergast, *The German Submarine War, 1914–1918* (London: Constable, 1931)

Glanfield, John, *The Devil's Chariots: The Birth and Secret Battles of the First Tanks* (Stroud: Sutton Publishing, 2006)

Goldrick, James, 'Wemyss, Rosslyn Erskine, Baron Wester Wemyss (1864–1933)', in *Oxford Dictionary of National Biography* (Oxford: Oxford University Press, 2004)

—— 'The need for a new naval history of the First World War', *Corbett Paper No. 7* (Corbett Centre for Maritime Policy Studies, 2011)

Gordon, G.A.H., *The Rules of the Game: Jutland and British Naval Command* (London: John Murray, 1996)

Goulter, Christina J.M., 'The Royal Naval Air Service: a very modern force', in Sebastian Cox and Peter Gray (eds), *Air Power History: Turning Points from Kitty Hawk to Kosovo* (London: Frank Cass, 2002)

Grainger, John D. (ed.), *The Maritime Blockade of Germany in the Great War: The Northern Patrol, 1914–1918* (Aldershot: Ashgate/Navy Records Society, 2003)

Gray, Edwin, *A Damned Un-English Weapon: The Story of British Submarine Warfare* (London: NEL, 1973)

Grey, Viscount, of Fallodon, *Twenty-Five Years: 1892–1916* (London: 1925)

Hallam, Richard and Mark Beynon (eds), *Scrimgeour's Small Scribbling Diary, 1914 – 1916: The Truly Astonishing Wartime Diary and Letters of an Edwardian Gentleman, Naval Officer, Boy and Son* (London: Conway, 2009)

Halpern, Paul G., *The Royal Navy in the Mediterranean, 1915–1918* (Aldershot: Navy Records Society, 1987)

—— *A Naval History of World War I* (London: UCL Press, 1994)

Halpern, Paul G. (ed.), *The Keyes Papers: Selections from the Private and Official Correspondence of Admiral of the Fleet Baron Keyes of Zeebrugge. Vol. II: 1919–1938* (London: Navy Records Society/George Allen and Unwin, 1980)

Hamilton, C.I., *The Making of the Modern Admiralty: British Naval Policy-Making, 1805–1927* (Cambridge: Cambridge University Press, 2011)

Hankey, Lord, *The Supreme Command, 1914–1918* (London: George Allen and Unwin, 1961)

Harding, Richard, 'The long shadow of the Dardanelles on amphibious warfare in the Royal Navy', in Peter Hore (ed.), *Dreadnought to Daring: 100 Years of Comment, Controversy and Debate in the Naval Review* (Barnsley: Seaforth, 2012), pp. 83–94

Harper, J.E.T., *The Truth about Jutland* (London: John Murray, 1927)

Harris, J.P., *Douglas Haig and the First World War* (Cambridge: Cambridge University Press, 2008)

Harrold, Jane and Richard Porter, *Britannia Royal Naval College, Dartmouth: An Illustrated History* (Dartmouth: Richard Webb, 2005)

Hart, Peter, *1918: A Very British Victory* (London: Weidenfeld & Nicolson, 2008)

von Hase, Georg, *Kiel & Jutland*, trans. Arthur Chambers and F.A. Holt (London: Skeffington & Son, n.d.)

Heathcote, T.A., *The British Admirals of the Fleet, 1734–1995: A Biographical Dictionary* (Barnsley: Leo Cooper 2002)

Hickey, Michael, 'Gallipoli: the Constantinople Expeditionary Force, April 1915', in T.T.A. Lovering (ed.), *Amphibious Assault: Manoeuvre from the Sea* (Royal Navy: Crown Copyright, 2005), pp. 2–22.

Hobbs, David, *Aircraft Carriers of the Royal and Commonwealth Navies* (London: Greenhill Books, 1996)

—— 'The first Pearl Harbour: the attack by British torpedo planes on the German High Seas Fleet planned for 1918', in John Jordan (ed.), *Warship 2007* (London: Conway, 2007)

Hodges, Peter, *The Big Gun: Battleship Main Armament, 1860–1945* (London: Conway Maritime Press, 1981)

Holloway, S.M. (ed.), *From Trench and Turret: Royal Marines' Letters and Diaries, 1914–1918* (London: Constable and Robinson, 2006)

Hopwood, R.A., *The Old Way and Other Poems* (London: John Murray, 1916)

—— *The New Navy and Other Poems* (London: John Murray, 1919)

Hore, Peter (ed), *Dreadnought to Daring: 100 Years of Comment, Controversy and Debate in The Naval Review* (Barnsley: Seaforth, 2012)

Hough, Richard, *Dreadnought: A History of the Modern Battleship* (London: George Allen and Unwin, 1968)

—— *The Great War at Sea, 1914–1918* (Oxford: Oxford University Press, 1983)

Howard, Michael, *The Continental Commitment: The Dilemma of British Defence Policy in the Era of Two World Wars* (Harmondsworth: Pelican Books, 1974)

—— *War in European History* (Oxford: Oxford University Press, 1976)

Hoy, Hugh Cleland, *40 O.B., or How the War Was Won* (London: Hutchinson, 1932)

Hunt, Barry D., *Sailor–Scholar: Admiral Sir Herbert Richmond, 1871–1946* (Waterloo: Wilfrid Laurier University Press, 1982)

Jackson, A.J., *De Havilland Aircraft since 1915* (London: Putnam, 1962)

—— *Blackburn Aircraft since 1909* (London: Putnam, 1968)

Jackson, Robert, *Strike from the Sea: A Survey of British Naval Air Operations, 1909–1969* (London: Arthur Barker, 1970)

Jane, Fred T., *Jane's Fighting Ships of World War I* (London: Studio Editions, 1990)

Jerrold, Douglas, *The Royal Naval Division* (n.l.: Naval and Military Press, 1923)

Johnston, Ian, Brian Newman and Ian Buxton, 'Building the Grand Fleet', in John Jordan (ed.), *Warship 2012* (London: Conway, 2012), pp. 8–21.

Kahn, David, *The Codebreakers* (London: Sphere Books, 1977)

Keble Chatterton, E., *The Big Blockade* (London: Hurst and Blackett, 1932)

—— *Danger Zone: The Story of the Queenstown Command* (London: Rich & Cowan Ltd, 1934)

Kemp, P.K. (ed.), *The Papers of Admiral Sir John Fisher* (n.l.: Navy Records Society, 1960, 1964)

Kent, Barrie, *Signal! A History of Signalling in the Royal Navy* (Clanfield: Hyden House, 1993)

Keyes, Sir Roger, *The Naval Memoirs of Admiral of the Fleet Sir Roger Keyes: The Narrow Seas to the Dardanelles, 1910–1915* (London: Thornton Butterworth, 1934)

Knight, Roger, *Britain against Napoleon: The Organization of Victory, 1793–1815* (London: Allen Lane, 2013)

Lambert, Andrew, *Admirals: The Naval Commanders Who Made Britain Great* (London: Faber and Faber, 2008)

—— *The Challenge: America, Britain and the War of 1812* (London: Faber and Faber, 2012)

Lambert, Nicholas A., *The Submarine Service, 1900–1918* (Aldershot: Ashgate / Navy Records Society, 2001)

—— *Sir John Fisher's Naval Revolution* (Columbia, SC: University of South Carolina Press, 2001)

—— *Planning Armageddon: British Economic Warfare and the First World War* (Cambridge, MA: Harvard University Press, 2012)

Lavery, Brian, *Able Seamen: The Lower Deck of the Royal Navy, 1850–1939* (London: Conway, 2011)

Layman, R.D., *Naval Aviation in the First World War: Its Impact and Influence* (n.l.: Caxton, 2002)

Llewellyn-Jones, Malcolm, *The Royal Navy and Anti-Submarine Warfare, 1917–49* (Abingdon: Routledge, 2006)

Lloyd, Christopher, *The Nation and the Navy: A History of Naval Life and Policy* (London: The Cresset Press, 1965)

Lloyd George, David, *War Memoirs of David Lloyd George* (London: Odhams Press, 1933–34, 1936)

Lovering, Tristan T.A. (ed.), *Amphibious Assault: Manoeuvre from the Sea – Amphibious Operations from the Last Century* (Royal Navy: Crown Copyright, 2005)

'Lower Deck', *The British Navy from Within* (London: Hodder & Stoughton, 1916)

Lowry, R.G., *The Origins of Some Naval Terms and Customs* (London: Sampson Low, Marston & Co., n.d.)

Mahan, Alfred Thayer, *The Influence of Sea Power upon History, 1660–1783* (London: Sampson Low, Marston & Co., 1890)

—— *The Influence of Sea Power upon the French Revolution and Empire 1793–1812*, 2 vols. (London: Sampson Low, Marston, Searle & Rivington, 1892)

Marder, Arthur J., *Portrait of an Admiral: The Life and Papers of Sir Herbert Richmond* (London: Jonathan Cape, 1952)

—— *From the Dreadnought to Scapa Flow: The Royal Navy in the Fisher Era, 1904–1919,* 5 vols. (London: Oxford University Press, 1961–70)

Massie, Robert K., *Dreadnought: Britain, Germany and the Coming of the Great War* (London: Pimlico, 1993)

—— *Castles of Steel: Britain, Germany and the Winning of the Great War at Sea* (London: Jonathan Cape, 2003)

Messimer, Dwight R., *Find and Destroy: Antisubmarine Warfare in World War I* (Annapolis, MD: Naval Institute Press, 2001)

Mountevans, Lord, *Adventurous Life* (London: Hutchinson, 1946)

Murfett Malcolm H. (ed.), *The First Sea Lords: From Fisher to Mountbatten* (Westport, CT: Praeger, 1995)

Newbolt, Henry, *Naval Operations: History of the Great War Based on Official Documents* (Uckfield: Naval and Military Press, n.d.)

North, John, *Gallipoli: The Fading Vision* (London: Faber and Faber, 1966)

Page, Christopher, *Command in the Royal Naval Division: A Miltary Biography of Brigadier General A.M. Asquith, DSO* (Staplehurst: Spellmount, 1999)

Palmer, Michael A., *Command at Sea: Naval Command and Control since the Sixteenth Century* (Cambridge, MA: Harvard University Press, 2005)

Parkes, Oscar, *British Battleships: Warrior 1860 to Vanguard 1950. A History of Design, Construction and Armament* (London: Leo Cooper, 1990)

Partridge, Michael, *The Royal Naval College Osborne: A History, 1903–21* (Stroud/Portsmouth: Sutton Publishing/The Royal Naval Museum, 1999)

Plunkett, Hon R., *The Modern Officer of the Watch* (Portsmouth: Gieves; London: John Hogg, 1913).

Ranft, Bryan, 'The protection of British seaborne trade and the development of systematic planning for war, 1860–1906', in Bryan Ranft (ed.), *Technical Change and British Naval Policy, 1860–1906* (London: Hodder & Stoughton, 1977)

Ranft, Bryan (ed.), *The Beatty Papers: Selections from the Private and Official Correspondence of Admiral of the Fleet Earl Beatty. Vol. 1: 1902–1918* (Aldershot: Navy Records Society/ Scolar Press, 1989)

Rankin, Nicholas, *Churchill's Wizards: The British Genius for Deception, 1914–1945* (London: Faber and Faber 2008)

Rasor, Eugene L., *British Naval History since 1815: A Guide to the Literature* (New York: Garland, 1990)

Reader, W.J., *Architect of Air Power: The Life of the First Viscount Weir of Eastwood, 1877–1959* (London: Collins, 1968)

Richmond, Sir H.W., *Command and Discipline* (London: Edward Stanford, 1927)

—— *Naval Training* (London: Oxford University Press/Humphrey Milford, 1933)

Roberts, John, *Battlecruisers* (London: Chatham Publishing, 1997)

Rodger, N.A.M., *The Admiralty* (Lavenham: Terence Dalton Ltd, 1979)

——— 'The Royal Navy in the era of the world wars: was it fit for purpose?' *Mariners Mirror* 97/1 (2011), pp. 272–84.

Roskill, Stephen, *H.M.S. Warspite: The Story of a Famous Battleship* (London: Collins, 1957)

——— *Hankey: Man of Secrets*, 3 vols. (London: Collins, 1970, 1972, 1974)

——— *Churchill and the Admirals* (London: Collins, 1977)

——— *Admiral of the Fleet Earl Beatty, The Last Naval Hero: An Intimate Biography* (London: William Collins, 1980)

Roskill, Stephen (ed.), *Documents Relating to the Naval Air Service. Volume 1: 1908–1918* (Bromley: Navy Records Society, 1969)

Schurman, Donald M. (ed.), *The Education of a Navy: The Development of British Naval Strategic Thought, 1867–1914* (Malabar, FL: Robert E. Krieger Publishing Company, 1984)

Seligmann, Matthew S., *The Royal Navy and the German Threat, 1901–1914: Admiralty Plans to Protect British Trade in a War Against Germany* (Oxford: Oxford University Press, 2012)

Sheffield, Gary, *Forgotten Victory: The First World War: Myths and Realities* (London: Review, 2002)

——— *The Chief: Douglas Haig and the British Army* (London: Aurum Press, 2011)

Shrub, R.E.A. and A.B. Sainsbury (eds), *The Royal Navy Day by Day* (Fontwell: Centaur Press, 1979)

Sumida, Jon Tetsuro (ed.), *The Pollen Papers: The Privately Circulated Printed Works of Arthur Hungerford Pollen* (London: Navy Records Society/George Allen & Unwin, 1984)

——— *In Defence of Naval Supremacy: Finance, Technology and British Naval Policy, 1889–1914* (Boston: Unwin Hyman, 1989)

——— *Inventing Grand Strategy and Teaching Command: The Classic Works of Alfred Thayer Mahan Reconsidered* (Washington, DC: Woodrow Wilson Center Press, 1997)

Temple Patterson, A. (ed.), *The Jellicoe Papers: Selections from the Private and Official Correspondence of Admiral of the Fleet Earl Jellicoe. Volume II: 1916–1935* (Greenwich: Navy Records Society, 1968)

——— *Jellicoe: A Biography* (London: Macmillan and Co., 1969)

Terraine, John, *Douglas Haig: The Educated Soldier* (London: Hutchinson, 1963)

——— *Business in Great Waters: The U-Boat Wars, 1916–1945* (London: Leo Cooper, 1989)

——— *White Heat: The New Warfare, 1914–18* (London: Leo Cooper, 1992)

Treadwell, Terry C., *The First Naval Air War* (Stroud: Tempus Publishing Ltd, 2002)

Tuchman, Barbara, *The Guns of August* (New York: Ballantine Books, 1994)

Warlow, Ben, *Battle Honours of the Royal Navy: Being the Officially Authorised and Complete Listing of Battle Honours Awarded to Her/His Majesty's Ships and Squadrons of the Fleet Air Arm Including Honours Awarded to Royal Fleet Auxiliary Ships and Merchant Vessels* (Liskeard: Maritime Books, 2004)

Wells, John, *The Royal Navy: An Illustrated Social History, 1870–1982* (Stroud: Alan Sutton Publishing, 1994)

Willis, Sam, *Fighting Ships, 1850–1950* (London: Quercus, 2008)

Wilson, Alastair and Joseph F. Callo, *Who's Who in Naval History: From 1550 to the Present* (London: Routledge, 2004)

Winseck, Dwayne R. and Robert M. Pike, *Communication and Empire: Media, Markets, and Globalization, 1860–1930* (Durham, NC: Duke University Press, 2007)

Winton, John, *Jellicoe* (London: Michael Joseph, 1981)

Wragg, David, *Fisher: The Admiral Who Reinvented the Royal Navy* (Stroud: The History Press, 2009)

Index

References in *italics* refer to illustrations or captions.